How to Be a Cheap Hawk

BROOKINGS STUDIES IN FOREIGN POLICY

The Brookings Institution is a private nonprofit organization devoted to research, education, and publication on important issues of domestic and foreign policy. Its principal purpose is to bring knowledge to bear on the major policy problems facing the American people.

On occasion Brookings authors produce relatively short studies that warrant immediate and broad circulation as contributions to public understanding of issues of current national importance. The Brookings Studies in Foreign Policy series is intended to make such studies available to a broad, general audience. In keeping with their purpose, these studies are not subjected to all of the formal review and verification procedures established for the Institution's research publications. As in all Brookings publications, the judgments, conclusions, and recommendations presented in the studies are solely those of the authors and should not be attributed to the trustees, officers, or other staff members of the Institution.

Studies in Foreign Policy

How to Be a Cheap Hawk

*The 1999 and 2000
Defense Budgets*

MICHAEL O'HANLON

BROOKINGS INSTITUTION PRESS
Washington, D.C.

Library of Congress Cataloging-in-Publication Data

O'Hanlon, Michael E.
    How to be a cheap hawk : the 1999 and 2000 defense budgets /
by Michael O'Hanlon.
        p.   cm. — (Brookings studies in foreign policy; 2)
    Includes bibliographical references and index.
    ISBN 0-8157-6443-X (pbk.)
    1. United States. Dept. of Defense—Appropriations and expendi-
tures.   2. United States—Military policy.   I. Title.   II. Series.
    UA23.6 .O33 1998
    355.6'229'0973—dc21   98-8940
                                                            CIP

                9 8 7 6 5 4 3 2 1

The paper used in this publication meets the minimum requirements
of the American National Standard for Information Sciences—
Permanence of Paper for Printed Library Materials, ANSI Z39.48-1984.

Typeset in Hiroshige and Copperplate

Composition by Harlowe Typography, Inc.,
Cottage City, Maryland

Printed by Kirby Lithographic
Arlington, Virginia

FOREWORD

What is the U.S. military's role in the post–cold war world? At first glance, one might think the issue has been largely settled. Congress and the president recently signed a balanced budget deal in which defense was among the least contentious issues. At virtually the same time, in mid-1997, the Clinton administration completed its Quadrennial Defense Review, which sustained much more than it changed about existing U.S. defense policy; notably, it continued to focus on global engagement and deterring war in both Northeast and Southwest Asia.

But big issues are lurking just below the surface of this apparent consensus on national military strategy and priorities. In testimony several months after the signing of the balanced budget deal, the Speaker of the House called attention to these issues by suggesting that any federal budget surpluses might be directed in large part to the Pentagon. A congressionally mandated National Defense Panel recently released a report suggesting that future battles will be unlikely to resemble conflicts like Desert Storm and that the military needs to prepare itself for that possibility. Uncertainty continues to reign over matters such as how robust to make NATO's military support for the security of any new members in Central Europe, how much to worry about China as a potential future military competitor, and how frequently to use the U.S. military to support peace operations or humanitarian missions like those conducted in Bosnia, Haiti, and Somalia. The budget picture, however, remains cloudy. Despite the attempts by the Quadrennial Defense Review to make Pentagon plans and programs fit within the five-year budget projections, the Defense Department may still have a force that is too expensive, given the fiscal constraints under which it has been directed to plan.

In this defense budget study, the latest in a long line of such Brookings monographs, Michael O'Hanlon wrestles with these and other challenges facing the Department of Defense today. He agrees with most of the main strategic priorities of the Clinton administration's Quadrennial Defense Review, but challenges its conclusions in a number of specific areas, including its affordability, its basic strategic construct of two major wars, and its plans for modernizing weaponry. Specifically, he further develops and defends his alternative to current U.S. military strategy: a slightly smaller force structure designed to undertake simultaneous "Desert Storm plus Desert Shield plus Bosnia peacekeeping" operations.

O'Hanlon is grateful especially for the assistance of Julien Hartley and Nick Boski with many of the research materials used in the study and the graphs and tables found therein. He also owes a particular debt of gratitude to Thomas McNaugher and David Mosher. Thanks are also due to Richard Betts, Stephen Biddle, Bruce Blair, Michael Brown, Robin Buckelew, Roxanne Cheney, Kurt Cichowski, Owen Cote, Robert Crumplar, Tom Davis, Randy DeValk, Tom Donnelly, Richard Dunn, William Durch, Joshua Epstein, Niclas Ericsson, Harold Feiveson, Ken Flamm, Aaron Friedberg, Michael Green, Susan Hardesty, John Hillen, Stuart Johnson, Scott Kennedy, Payne Kilbourn, Lawrence Korb, Richard Kugler, Nick Lardy, Kaori Lindeman, Frances Lussier, Terrence Lyons, Paul McHale, Robert McNamara, David Meade, Mike Mochizuki, Janne Nolan, Eric Nyberg, David Ochmanek, George Reed, Steve Sargeant, Stephen Schwartz, Stephen Solarz, Paul Stares, Stephen Stedman, John Steinbruner, Frank von Hippel, Susan Woodward, and a number of civilian and uniformed employees of the Department of Defense. The project was performed under the direction of Richard Haass.

Kerry V. Kern edited the manuscript, Ellen Garshick proofread the pages, and Susan M. Fels provided the index.

Brookings gratefully acknowledges the financial support of the Carnegie Corporation of New York and the John D. and Catherine T. MacArthur Foundation.

The views in this study are those of this author alone and should not be attributed to those acknowledged above or to the trustees, officers, or other staff members of the Brookings Institution.

MICHAEL H. ARMACOST
President

March 1998
Washington, D.C.

CONTENTS

CONTENTS

FIGURES

1
OVERVIEW AND SUMMARY

Once again, the Pentagon has recently completed a major strategic and budgetary analysis. The 1997 Quadrennial Defense Review (QDR) was completed in May under the guidance of Secretary of Defense William S. Cohen. It was later fully fleshed out in the November 1997 Defense Reform Initiative, which detailed some of the personnel cuts recommended by the QDR, and in President Clinton's budget submission for fiscal year 1999.

Unlike the decade's earlier defense reassessments, which led to the Bush administration's Base Force concept in 1991 and the Clinton administration's Bottom-Up Review (BUR) in 1993, the QDR was not occasioned by major realignments on the world scene or at the White House. For those reasons, it continues and reinforces more about U.S. security policy than it changes. Virtually all major force structure elements—Army divisions, Air Force wings, and so forth—remain unchanged by the QDR, and no major weapons procurement program is fundamentally altered or canceled. Emphases on maintaining global military presence, preparing for two overlapping wars that each could resemble Desert Storm in character, deploying a large nuclear force, and devoting large sums of money to military research and innovation are preserved.

Secretary Cohen's strategic doctrine does make some useful adjustments, however. Notably, it places greater emphasis on defending deployed troops as well as the American homeland against weapons of mass destruction. Its declaratory strategy also places greater emphasis on trying to "halt" an enemy attack in its early phases and on conducting peace operations that save lives and help preserve America's moral and political claim to global leadership.[1]

The QDR also shows how annual defense spending requirements can be reduced by about $10 billion by the early years of the next decade without scaling back the U.S. role in the world. It would realize these savings through cuts in full-time defense employment of about 5 percent and reductions in planned purchases of several major weapons systems typically amounting to about 20 percent of the previously anticipated quantities. To shave that much money off the defense budget without significantly impairing military capability is no mean feat. Although still considerable at $265 billion in 1998, U.S. defense spending is already down $100 billion from the 1980s average and $50 billion from the overall cold war level.

However, the QDR also merits criticism. Most notably, though it took a modest step in the right direction, it failed to eliminate a looming funding shortfall at the Defense Department. The gap between planned spending and military requirements will arise principally because of the need to increase equipment spending to replace or "recapitalize" an aging force. That need will become acute even as real defense spending levels continue to decline modestly through 2002. Even if the Pentagon obtains the real spending increase in 2003 projected in President Clinton's 1999 budget request, procurement spending that year will remain about $10 billion shy of the annual level that will be needed to sustain the QDR force over time. Also, tight funds will make it difficult to contemplate new expenditures— such as those for additional peace operations or, should it prove technically feasible, some type of national missile defense system.

The QDR also failed to wrestle deeply with issues about the future of U.S. alliances and global security institutions. It tacitly approved an undesirable situation in which major U.S. allies remain generally unprepared for conflicts that could threaten common Western interests in places like the Persian Gulf. That may be more a shortcoming of security strategists at the White House and State Department than

1. For a more critical view, see Michael G. Vickers and Steven M. Kosiak, *The Quadrennial Defense Review: An Assessment* (Washington: Center for Strategic and Budgetary Assessments, 1997).

of defense experts, who would clearly be imprudent to incorporate into their war plans allied capabilities that do not now exist. It remains a subject of acute interest to the Pentagon, however, and should have received more consideration.

Although it was originally intended to provide a constructive critique of the QDR and suggest ways to remedy its flaws, the report of the National Defense Panel (NDP) repeats most of these mistakes. The report does make some useful contributions. It wisely challenges the assumption that U.S. forces will be able to safely establish large beachheads from which to wage future battles the way they did in Desert Storm. The panel advocates making ground units more mobile and dispersed in battle, fully exploiting unmanned aerial vehicles and other new reconnaissance and communications systems, conducting more joint-service experimentation using new technologies and operational concepts, and making greater efforts to protect deployed troops as well as the American homeland from attacks involving weapons of mass destruction. The latter recommendation, for example, appears to have helped convince the Pentagon to dedicate $50 million in its 1999 budget request for civil preparedness against such attacks and to propose an increase in the cooperative threat reduction (or Nunn-Lugar) program from the 1998 level of $382 million to $464 million.[2]

The NDP report avoided tackling the majority of difficult issues head on, however. Its most important shortcomings are on the budgetary front. The few specific proposals for saving money—notably its suggestion that no more M1A1 Abrams battle tanks be upgraded to the A2 configuration, that the final Nimitz-class aircraft carrier not be built, and that more defense support activities be competed—are rather modest in budgetary scope. Taken together, they are unlikely to redress more than $1 billion to $2 billion of the annual funding shortfall that will likely arise next decade under the QDR force. Most of the report's other suggestions are vague, as in its observation that the two-war strategy may be outliving its usefulness and in its "questions" about whether the Pentagon should purchase the total number of fixed-wing and rotary-wing combat air-

2. Bryan Bender, "DoD Approves Guard, Reserve Role in Domestic Terrorism," *Defense Daily*, January 28, 1998, p. 5; and Department of Defense, "FY 1999 Defense Budget Briefing," February 2, 1998.

craft now planned. Meanwhile it calls for an added $5 billion to $10 billion in annual funding to support initiatives in areas like intelligence, space, urban warfare, joint experimentation, and information operations as part of a "transformation strategy." To the extent that such initiatives were added to the existing Pentagon plan, and only modest or token cuts made in programs that the NDP considers excessive, the budgetary shortfall of the DoD could actually get worse than now projected.[3]

THE U.S. ROLE IN THE WORLD AND THE QDR

Current U.S. security posture defies simple characterization because it responds to a complex international environment containing a number of mid-level and disparate threats. Some could intensify with time; others are likely to recede. Some are more amenable to being addressed through the judicious use of military force than others. In this context, it is not surprising that the U.S. security community has failed to settle on a single unifying slogan to characterize post–cold war policy. If there were such a widely accepted buzzword or pithy description, it would probably be misleading, if not dangerous.

To deal with this uncertain environment, U.S. security policy relies on a wide variety of formal and informal structures with other countries. They include formal alliances with Japan, South Korea, Australia, New Zealand, the Philippines, Thailand, most of western Europe, and nearly all the countries in the Western Hemisphere; ad hoc coalitions, often under UN imprimatur; and arms control treaties with a wide variety of the world's countries including Russia and China.

The core commitments of the United States are to the stability of the North American region and the Eurasian littoral from Europe to the Persian Gulf and into Japan and Korea. But its grand strategy is also founded more broadly on a recognition that a world involving weapons of mass destruction, shifting distributions of wealth and power, and major demographic, economic, and environmental interdependencies is too dangerous to withdraw from.[4] Such a world can-

3. National Defense Panel, *Transforming Defense: National Security in the 21st Century* (Arlington, Va.: December 1997), pp. 23, 49, 59, 79–86.

4. Such a depiction of the character of the future global security environment is not a fuzzyheaded, liberal vision; the Defense Intelligence Agency makes these

not, as Richard Haass argues, be left "unregulated."[5] Indeed, the degree to which U.S. leadership and post–World War II institutions already strengthen and stabilize it should not be underestimated.[6] This approach should probably be expanded over time by developing security relationships with other countries as conditions allow.[7]

Overall demands on U.S. military forces today are clearly less than in the cold war. There is no peer rival to the United States today. Major structural economic problems facing Russia and China would have to be solved before those nations could pose a major military challenge to U.S. interests even if they wanted to.[8] But other challenges remain significant. They include deterring radical regimes in North Korea and Iraq, as well as conflict in the Taiwan Straits; undergirding NATO and U.S.-Asian alliances; keeping a lid on proliferation of weapons of mass destruction, as well as hedging against the real possibility of their use; and trying to expand what Richard Ullman of Princeton University calls the "zone of peace" to regions where conflict and instability are enduring conditions.[9]

The QDR's unglamorous but still apt description for this approach to U.S. security policy is "shape, respond, prepare."[10] That is, its

---

issues its first order of business in its 1997 assessment of possible threats to U.S. security. See Lieutenant General Patrick M. Hughes, Defense Intelligence Agency, "Global Threats and Challenges to the United States and Its Interests Abroad," statement for the Senate Select Committee on Intelligence, February 5, 1997, pp. 1–10.

5. Richard N. Haass, *The Reluctant Sheriff: The United States after the Cold War* (Washington: Council on Foreign Relations, 1997), pp. 21–74.

6. See G. John Ikenberry, "The Myth of Post–Cold War Chaos," *Foreign Affairs*, vol. 75 (May–June 1996), pp. 79–91.

7. Ashton B. Carter, William J. Perry, and John D. Steinbruner, *A New Concept of Cooperative Security* (Brookings, 1992); and Janne Nolan, ed., *Global Engagement: Cooperation and Security in the 21st Century* (Brookings, 1994), pp. 3–61.

8. For a sober assessment of China's deep economic problems, see interview with Nicholas Lardy, "Ailing Chinese Bank System Brings Pressure to Privatize," *Washington Times*, September 26, 1997, p. A14.

9. Richard H. Ullman, "Enlarging the Zone of Peace," *Foreign Policy*, no. 80 (Fall 1990), pp. 102–20.

10. These concepts also appear in the most recent White House security document; see Bill Clinton, *A National Security Strategy for a New Century* (May 1997), pp. 6–14.

underlying strategy seeks to shape the character of the international environment through forward presence, military-to-military exchanges, military education programs, joint exercises, and security assistance; to respond to threats, through forward presence and other instruments of deterrence and with military force, if necessary; and to prepare for an uncertain future through technological, tactical, and operational innovations in U.S. armed forces.[11] Its explicit emphasis on a wide range of military tools and missions is more realistic than the Bottom-Up Review's preoccupation, at least in theory if not in practice, with the major regional war scenario.[12]

The Clinton administration has often described its strategy as one of engagement and enlargement. That description understates the hard-headed deterrent aspects of current U.S. security policy. It also says little about how to enlarge the community of democracies in practice.[13] But it does properly emphasize the fact that, as far as we can tell, modern, mature, and prosperous liberal democracies seem unlikely to go to war with each other. It provides a vision that, if applied pragmatically and carefully, can also reinforce America's right in the eyes of other countries to act the part of the world's only superpower. It could also provide some integrating vision to relations with countries that are not currently democratic, such as China.[14]

THE MAIN ELEMENTS OF THE QDR

In light of this complex international environment and the multiple demands it places on U.S. armed forces, the QDR was wise to pre-

11. William S. Cohen, *Report of the Quadrennial Defense Review* (Department of Defense, May 1997), pp. 25–26.

12. Notably, the Bottom-Up Review document considered activities like peacekeeping and other operations other than war to be "lesser cases" that could simply be handled with a subset of the warfighting instruments. See Les Aspin, *Report on the Bottom-Up Review* (Department of Defense, October 1993), p. 23.

13. Richard N. Haass, "Fatal Distraction: Bill Clinton's Foreign Policy," *Foreign Policy*, no. 108 (Fall 1997), p. 113.

14. As one well-known realist, Richard Betts of Columbia University, puts it in regard to China, "We may begin with a realist diagnosis but be forced into banking on liberal solutions, simply because the costs of controlling the balance of power may be too high." See Richard K. Betts, "Wealth, Power, and Instability: East Asia and the United States after the Cold War," *International Security*, vol. 18 (Winter 1993–94), p. 55.

serve a range of U.S. capabilities and defense priorities. But the specifics of its approach can and should be challenged. The concrete military pillars of the QDR include the following:

—First, being ready for the two-major-war scenario as currently understood, with overlapping regional conflicts resembling Desert Storm in scale and basic characteristics;

—Second, conducting the global presence mission, consisting of naval deployments in the western Pacific Ocean, Mediterranean Sea, and Persian Gulf region, as well as strong Army and Air Force postures in Japan, Korea, and Western Europe;

—Third, maintaining a nuclear force posture with 3,500 strategic warheads (though more until the START II Treaty is ratified by the Russian Duma), and with significant attention to countering weapons of mass destruction through arms control, defenses, and preemptive capabilities of one type or another; and

—Fourth, modernizing U.S. weaponry to maintain substantial technological supremacy over any possible foe and redress specific weaknesses in U.S. forces.

Turning from the general to the specific, the QDR outlines a number of concrete changes in U.S. armed forces. In the realm of personnel policy, it envisions reducing active-duty uniformed troops by roughly 60,000 people, to a total of 1.36 million, probably saving about $3 billion annually. Those cuts would include 15,000 Army troops, 18,000 sailors, 26,000 uniformed Air Force personnel, and 1,800 Marines. (The services' total uniformed strengths would then be approximately 480,000 for the Army, 369,000 for the Navy, 339,000 for the Air Force, and 172,000 for the Marines—making for respective reductions since 1990 of 36 percent, 37 percent, 37 percent, and 13 percent).[15]

Additional savings would result from other personnel reductions. Full-time civilian employees are to decline by 80,000, to 640,000—saving about $4 billion a year (though new costs will be incurred due

15. The respective declines of the four services' sizes vary slightly depending on the initial point of comparison. But regardless of the recent benchmark chosen, be it 1980 or 1985 or 1990 or some other recent year, post–cold war cuts in uniformed personnel in the three main services have been relatively equal to each other in proportion. See William S. Cohen, *Annual Report to the President and the Congress* (Department of Defense, April 1997), p. C-1; and Department of Defense, "FY 1999 Defense Budget Briefing," February 2, 1998.

to outsourcing). Cuts in the military reserve component would total 55,000 individuals, resulting in a total of 835,000 troops in the selected reserves. That will reduce annual spending by nearly a billion additional dollars. Most of those reductions—about 38,000 reservists—will apparently be in the Army National Guard.[16] By 2003, the reductions in troops will be nearly completed (active-duty forces will remain 6,000 greater than the QDR goal, and reservists 2,000 greater). Civilian employees of the Department of Defense will still number 32,000 more than the ultimate goal.[17]

The QDR also recognizes that DoD still maintains much too large an infrastructure given the size of its active-duty force. Numbers of military personnel are down by more than 30 percent, but the number of major domestic bases is being streamlined only by 20 percent. Although one can argue that it would be desirable to retain an infrastructure adequate for a larger force than today's in case of a future mobilization, it is more pressing that DoD have the funds to modernize the force and keep it ready. (At most, the Pentagon might hold on to some of the land where it shuts down bases—but even that measure is generally not needed or appropriate.) In a world characterized less by traditional great-power competition than by rogue actors, nonstate threats, and weapons of mass destruction, preparing for hypothetical mobilization scenarios is not as high a priority as near-term military readiness and technical and doctrinal innovation. The QDR therefore rightly recommended closing more bases—two more rounds, in 1999 and 2001, on top of the four undertaken since 1988—and privatizing various defense support functions. Doing so would save another $3 billion to $4 billion a year and presumably allow further personnel reductions (of unspecified magnitude).

In addition, Secretary Cohen scaled back planned production runs for the Air Force's F-22 and Navy's F/A-18 E/F combat aircraft by 22 percent and 45 percent respectively, freeing up another $2 billion a year in the next decade. He reduced the scheduled production of Joint STARS reconnaissance and targeting aircraft by six, reduced the planned purchase of V-22 Osprey tilt-rotor aircraft by 15 percent

16. Frances Lussier, *Structuring the Active and Reserve Army for the 21st Century* (Congressional Budget Office, December 1997), p. 29.

17. Department of Defense, "FY 1999 Defense Budget Briefing," February 2, 1998.

(though near-term purchases were increased slightly), and made several other modest changes in smaller weapons programs that together will save about $1 billion annually.[18]

Adding all these savings together, the QDR could produce annual savings exceeding $10 billion.[19] But given the state of play in early 1998—specifically Congress's reluctance to close more bases, reduce reservists, or allow the Pentagon more flexibility in how it maintains and repairs weapons—the QDR is on track to realize annual savings of less than $10 billion.

Hope should not be abandoned too soon, however. Secretary Cohen reiterated his call for more base closures—and by a more politically tenable schedule, with rounds now requested for 2001 and 2005—in his November 1997 Defense Reform Initiative Report. In that report he also fleshed out a number of the specifics of how the QDR's personnel cuts could be made, calling for the elimination of 30,000 positions in defense agencies, the office of the secretary of defense, and other headquarters. Only civilian jobs are actually to be eliminated (military personnel now in such jobs, by contrast, will be reassigned), meaning that the net reduction in employment is less than 30,000. It is still significant: 15,000 positions would translate into $750 million in annual savings. Adding in anticipated savings from elimination of most paperwork in defense acquisition practices (relying instead on computers, specialized credit cards for small purchases, and other innovations), increased use of competition to determine who will carry out defense support functions, and other reforms, the Pentagon hopes to achieve $6 billion in steady-state savings from the defense reform initiative. As noted, those savings are not over and above what had been anticipated by the QDR. They give further specificity—and, in the case of base closures, perhaps further political life—to ideas that are found in the QDR. That may in turn improve the odds that the entire 1997 defense restructuring process will ultimately deliver at least $10 billion in annual savings.[20]

18. Cohen, *Report of the Quadrennial Defense Review*, pp. vii–viii.

19. DoD arrives at an estimate of roughly $10 billion in expected savings, assuming that Congress grants the requested base closure authority. Cohen, *Report of the Quadrennial Defense Review*, pp. 61–62.

20. William S. Cohen, *Defense Reform Initiative Report* (Department of Defense, November 1997); and David A. Fulghum, "Pentagon Reforms Spark Concerns," *Aviation Week and Space Technology*, November 17, 1997, pp. 30–31.

The QDR also recommended slightly more than $1 billion a year in additional expenses. An annual average of about $400 million was added to the national missile defense research and development budget. That addition is intended to assure the integrity of the "3+3" process, whereby a decision to deploy a nationwide defense could be made as soon as 2000 and result in an operational system three years later. Nearly $200 million was added to the annual counterproliferation budget, to be focused on missions such as decontamination and protection from chemical weapons, detection of biological weapons, and greater support for special operations forces. Modest amounts of money were added for cruise missile defense and information warfare and protection. It might also be fair to include DoD's planned expenditures of perhaps $50 million to $100 million to assist the process of NATO expansion in this tally, since the NATO expansion policy was effectively developed between the BUR and QDR. Small amounts were also added to reduce the strain on certain "high-tempo/low-density" units like those flying AWACS planes by adding more crews or equipment for some tasks.[21]

Further initiatives of larger magnitude were unveiled with President Clinton's 1999 budget request. Funding for the Army's "digitization" efforts to improve data transmission between soldiers is to increase gradually from $2.6 billion in 1998 to $3.2 billion in 2003. Most notably in dollar terms, and repeating a pattern that has become common in the 1990s, several billion dollars were shifted from procurement accounts to operations accounts between the 1998 and 1999 budget requests. For the years 1999 through 2002, the 1999 budget request projects that budget authority for procurement will be $12 billion less, and funds for operations and maintenance $18 billion more, than envisioned in the 1998 request.[22] Adding funds to operating accounts, and providing a 3.1 percent pay raise for troops in 1999, are wise policies, as argued further in chapter 5—but that does not change the fact that procurement accounts remain insufficient to sustain the QDR force over the medium to long term.

21. Cohen, *Report of the Quadrennial Defense Review*, pp. 48–49, 61; and Department of Defense, "FY 1999 Defense Budget Briefing," February 2, 1998.

22. Department of Defense, "FY 1998 Defense Budget Backup Charts," February 6, 1997; and Office of the Assistant Secretary of Defense (Public Affairs), "Department of Defense Budget for FY 1999," February 2, 1998, p. 7.

THE QDR IN CONTEXT

How do these specific changes proposed by the Quadrennial Defense Review look in perspective? The defense spending context is examined in some detail in the next chapter; the following summarizes issues like force structure, overseas basing, and defense infrastructure.

Considering first the matter of military manpower, as of September 30, 1997, the United States had 1.45 million active-duty service members (see table 1-1); as of September 30, 1998, it is expected to have about 1.43 million. That latter number is almost down to the 1.42 million that had been called for in the 1993 Bottom-Up Review. The QDR anticipates additional reductions down to 1.36 million. As of early 1998, the post–cold war troop drawdown is about 90 percent complete. The reduction in full-time DoD civilian employees is about two-thirds complete.

By way of international comparison, the United States currently has the second largest armed forces in the world. China's remain the largest—indeed, at 3 million troops, its military is roughly twice the size of the U.S. armed forces, though President Jiang Zemin recently announced a further reduction of 500,000 troops to be effected over the next three years.[23] Russia's military, larger than that of the United States until 1995, is now just slightly smaller and is scheduled to remain that way under the recent reform plan of President Boris Yeltsin to cut active-duty troops to 1.3 million or less.[24] The only other militaries with more than 1 million active-duty troops are those of India and North Korea. Pakistan, Egypt, Iran, Iraq, and Syria each have about 400,000 to 500,000 troops.

Among major U.S. allies, South Korea's forces of about 600,000 are largest, Turkey's 500,000 are the next most numerous, and Italy, Germany, and France each typically field around 350,000 troops. But the smallest of the major European militaries, Britain's, is probably the continent's best all-around armed force for the likely challenges of the post–cold war world, just as Israel's active force of less than 200,000 is clearly the best in the Middle East.[25]

23. Robert Karniol, "PLA Force Strength Will Be Cut by Half a Million," *Jane's Defence Weekly*, September 24, 1997, p. 13.

24. See John Thornhill, "Yeltsin in War of Words over Military Cuts," *London Financial Times*, July 22, 1997, p. 2.

25. International Institute for Strategic Studies, *The Military Balance 1996/97* (Oxford, England: Oxford University Press, 1996), pp. 306–11.

TABLE 1-1. *Defense Manpower*

Thousands of troops unless otherwise indicated

| Component | BUR plan[a] | | | QDR plan | Percent change, 1990–97 | Percent change, planned, 1997–2003 |
	1990	1997	2003			
Active	2,069	1,450	1,420	1,360	−30	−6
Reserve	1,128	900	890	835	−20	−7
Civilian	1,070	800	720	640	−25	−20

Sources: William S. Cohen, *Report of the Quadrennial Defense Review* (Department of Defense, May 1997), p. 31; and William J. Perry, *Annual Report to the President and the Congress* (Department of Defense, February 1995), pp. 275, C-1.

a. Numbers for 1990, 1997, and 2003 are as of September 30.

With only a few minor exceptions, the U.S. military has now reached its anticipated post–cold war force posture—the QDR having no effect on most of it. As shown in table 1-2, active Army divisions now number ten; Navy aircraft carrier battle groups number eleven active and one reserve units along with ten active and one reserve air wings; Air Force fighter wings are now at their final anticipated level of twenty, although the mix between active and reserve is now to be changed from thirteen and seven to twelve and eight.

Some DoD analyses reportedly suggested that a two-major-war strategy could be ensured with somewhat smaller forces. Depending on the study, requirements were calculated to be as small as eight active-duty Army divisions, sixteen Air Force fighter wing equivalents, or ten Navy carrier battle groups (different studies reached different conclusions about which forces could be reduced). But the current force structure was maintained on the grounds that war plans appear safer with a military of that size. Specifically, the Pentagon argues that if adversaries used weapons of mass destruction, attacked with less warning than expected, or posed other major and unanticipated challenges, all of the major warfighting force posture and possibly a number of Army National Guard enhanced brigades might well be needed to prevail decisively.[26]

26. Cohen, *Report of the Quadrennial Defense Review*, p. 24; and U.S. Army Concepts Analysis Agency, "Strategy-Based Campaign Analysis," Draft Working Paper, Bethesda, Md., 1997.

TABLE 1-2. *Major Elements of Force Structure*

Service unit	BUR plan[a]			QDR plan
	1990	1997	2003	
Army				
Active divisions	18	10	10	10
Reserve brigades	57	42	42	30
Navy				
Aircraft carriers (active/reserve)	15/1	11/1	11/1	11/1
Air wings (active/reserve)	13/2	10/1	10/1	10/1
Attack submarines	91	73	52	50
Surface combatants	206	128	131	116
Air Force				
Active fighter wings	24	13	13	12+
Reserve fighter wings	12	7	7	8
Reserve air defense squadrons	14	10	6	4
Bombers (total)	277	202	187	187
Marine Corps				
Marine expeditionary forces	3	3	3	3

Sources: Cohen, *Report of the Quadrennial Defense Review*, pp. 30, 33; Perry, *Annual Report to the President and the Congress* (February 1995), pp. 275, D-1; (March 1996), p. D-1; and International Institute for Strategic Studies, *The Military Balance 1990–1991* (Macmillan Publishing, 1990), pp. 17–27.

a. Numbers for 1990, 1997, and 2003 are as of September 30.

Another reason for sustaining most of the BUR force structure was to avoid exacerbating a high tempo of operations that has strained the military and caused concerns over morale and troop retention. From Bosnia to Iraq to Somalia to Haiti, the post–cold war U.S. military has been quite busy; these missions, combined with the demands of normal training and exercises, have been difficult for the U.S. military to cope with. As a remedial step, the QDR also elected to reduce joint-service training exercises by at least 15 percent and perhaps as much as 25 percent, since they keep troops away from home base and are sometimes overused by regional military commanders (the "CINC's," or commanders-in-chief in the field).[27]

27. Speech of General John Shalikashvili, National Press Club, September 24, 1997.

Nuclear forces will be influenced much less by the QDR than by the START II ratification process. If START II is ratified by the Russian Duma, deployed strategic warheads will decline to a total of 3,500, including elimination of the MX or "Peacekeeper" missile, de-MIRV'ing of the Minuteman III force to 1 warhead per missile, definitive conversion of the B-1 bomber fleet to a strictly conventional role, and reduction in the Trident submarine fleet to 14 ships each with 5 warheads on 24 separate missiles. But until that treaty is ratified by the Russian Duma, U.S. forces are otherwise expected to remain more or less unchanged with around 8,000 long-range warheads. They will continue to be based on 50 MX missiles carrying 10 warheads each; 500 Minuteman III ICBMs carrying 3 warheads each; 18 Trident submarines, each with 24 missiles carrying 8 warheads for a total of 192 warheads per boat; and roughly 94, 71, and 21 B-1, B-52, and B-2 bombers respectively, carrying in turn up to 16, 20, and 16 warheads each (though the B-1 fleet appears to be increasingly dedicated to conventional missions even today).[28]

The major proposed force-structure effects of the QDR are to ratify the planned reductions in the Army National Guard's combat force structure, from roughly 42 brigade-equivalents to about 30;[29] to further reduce dedicated continental air defense interceptor aircraft by 2 squadrons; to maintain 50 attack submarines (the Bottom-Up Review had indicated a range of 45 to 55); and to reduce the total number of major surface combatants from 131 to 116 (including cruisers, destroyers, and frigates, of which there are now about 30, 50, and 50 in the fleet, respectively). These changes are significant. But they are also secondary in the sense that they do not effect the "core forces"— active Army divisions, Air Force fighter wings, aircraft carrier battle groups, and Marine units—that figured most prominently in the BUR's assumed combat package for major regional wars.

Defense infrastructure is also declining, as noted, though by less than the one-third reduction in active-duty troops and forces since

28. Cohen, *Report of the Quadrennial Defense Review*, p. 32; "Factfile," *Arms Control Today*, vol. 26 (October 1996), p. 28; and *Budget of the United States Government, Fiscal Year 1999*, p. 135.

29. It is possible that the number of Guard combat brigades could be further reduced, but that is not yet clear; see Lussier, *Structuring the Active and Reserve Army*, p. 29.

FIGURE 1-1. *Base Closures in Perspective*[a]

Percent

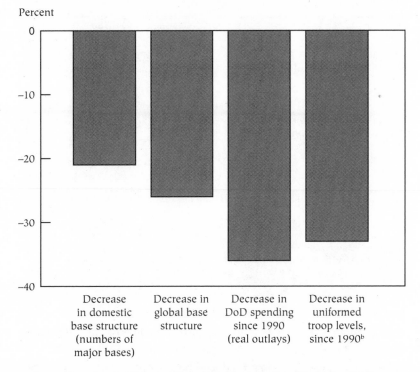

| | Decrease in domestic base structure (numbers of major bases) | Decrease in global base structure | Decrease in DoD spending since 1990 (real outlays) | Decrease in uniformed troop levels, since 1990[b] |

Source: Department of Defense, Office of Program Analysis and Evaluation, briefing.
a. For 2002, assuming completion of 1988, 1991, 1993, and 1995 BRAC rounds.
b. Spending and troop levels in 1990 were already slightly below their Reagan-era peaks.

1990 (see figure 1-1). The total number of major domestic bases will have declined by 97, from roughly 500 to 400, once the installations slated for closure by the Base Realignment and Closure Commissions of 1988, 1991, 1993, and 1995 are all completely shut down. Only the first two rounds have run their full course, while the 1993 round is about half completed and the 1995 round is just beginning to have an appreciable effect. In all, roughly 60 major domestic bases were closed by the end of fiscal year 1997. Net savings from the closures have been realized more slowly than had been hoped, but each round has been breaking even within about five years. Total annual savings from the four rounds combined are expected eventually to approach

TABLE 1-3. *U.S. Military Personnel in Foreign Areas, 1986–96*[a]

Thousands of troops

Country	1986	1991	1996
Germany	250	203	49
Other European countries	75	62	62[b]
Europe, afloat	33	20	4
South Korea	43	40	37
Japan	48	45	43
Other Asia-Pacific countries	17	9	1
Pacific afloat (including southeast Asia)	20	11	15
Latin America/Caribbean	13	19	12
Miscellaneous	26	39[c]	17
Total	525	448	240

Source: William S. Cohen, *Annual Report to the President and the Congress* (Department of Defense, April 1997), p. C-2.

a. As of September 30 of each year. Numbers may not add up to totals due to rounding.

b. Includes 26,000 troops in the former Republic of Yugoslavia and Hungary in support of operations in Bosnia and Herzegovina. (In 1995, before the IFOR deployment in Bosnia, the United States had about 37,000 troops in the "Other European countries" category.)

c. Does not include an additional 118,000 shore-based forces in support of Operation Desert Storm.

$6 billion.[30] The QDR's recommendations for two more rounds would, as noted above, add half again as many savings.

Overseas bases have been reduced by slightly more than half since the end of the cold war, roughly in keeping with the reductions in overseas military manpower indicated in table 1-3.[31] At present the United States maintains some 100,000 troops in the Asia-Pacific region and 100,000 in the European theater at any time. The former forces are found mainly in Japan and South Korea, with the remaining 15,000 to 20,000 afloat, on Guam, and in smaller numbers in a variety of countries. About half of the European forces are located in Germany, with roughly 15,000 in Italy, 10,000 in the United Kingdom, 13,000 distrib-

30. Wayne Glass, *Closing Military Bases: An Interim Assessment* (Congressional Budget Office, December 1996), pp. 2, 30, 32, 63–68; and Department of Defense, *Defense Almanac 96* (Alexandria, Va.: American Forces Information Service, 1996), pp. 48–49.

31. Glass, *Closing Military Bases*, pp. 2, 30, 32, 63–68; and Department of Defense, *Defense Almanac 96*, p. xi.

uted in Bosnia and Hungary, 3,000 in Spain, and several thousand at sea in the Mediterranean.[32]

NEXT STEPS

Where, if anywhere, should the Pentagon go from here? As noted, the QDR makes several adjustments to both U.S. security strategy and the specifics of the defense program and armed forces that are not unreasonable. But its budgetary realism is open to serious doubt. Because real defense spending is still headed downward at a time when most force structure cuts have already been completed and weapons spending accounts need bolstering—even without including any major new programs such as a national missile defense or "revolution in military affairs" (RMA) initiative—the Pentagon will soon face a significant gap between projected budgets and spending requirements.

How big is that shortfall likely to be? The question is actually difficult to answer. Future defense budget requirements are difficult to calculate precisely. Weapons often cost more than expected, unexpected burdens arise in areas like environmental cleanup and health care, and savings from various approaches to making the Pentagon more efficient often prove to be of far smaller size than is hoped. It is even more difficult to foresee the likely level of future defense resources. President Clinton's 1999 budget request projects a gradual decline of real defense spending over the period when the balanced budget deal's constraints are operative, from the 1998 level of $265 billion to a 2002 level of $250 billion (including the $10 billion or so the Department of Energy will be spending on nuclear weapons–related activities, and expressed in constant 1998 dollars, as are all figures in this study unless otherwise noted). Yet the budget request projects a real defense spending increase of $10 billion in the 2003 budget. That may or may not prove realistic; it does appear at odds with the president's promise in his January 1998 State of the Union address to withhold any future federal budget surplus until social security is restructured and protected, and also appears at odds with most other politicians' preferences for how to use any surplus.

32. Cohen, *Annual Report to the President and the Congress* (1997), p. C-2; and International Institute for Strategic Studies, *Military Balance 1996/97*, pp. 28–31.

My best estimate is that the QDR force in its present form would cost roughly $270 billion a year to sustain over the next decade. That is $10 billion above the projected 2003 level and $20 billion above the somewhat more politically tested 2002 level. So depending on the reference point chosen, the annual shortfall would amount to somewhere between $10 billion and $20 billion in the next decade.

This study outlines a possible solution to the Pentagon's budget problem. The basic thrust of my alternative warfighting strategy and force posture can be encapsulated in the following recommendations:

Modify the "two Desert Storm" approach to force sizing to something more akin to a "Desert Storm plus Desert Shield plus Bosnia" posture. Given the unlikelihood of nearly simultaneous conflicts, and the further improbability that if overlapping wars occurred the United States and regional allies would lose the initial battles and require huge U.S. reinforcements in both, that posture should provide solid deterrence. Also, given the devastating capabilities of modern airpower, even the Desert Shield–scale deployment would have considerable attacking power (though it would probably not suffice for a major ground counteroffensive). The corresponding economies should not be overstated, however: this strategic construct would require a force about 90 percent as large and expensive as under the QDR. In order to preserve enough units to maintain the high operations tempo that missions in places like Bosnia and Iraq have imposed on the U.S. armed forces, many of these cuts would best be made in the size of individual units such as Army and Marine divisions rather than in their number.

Change routine forward naval operations by rotating crews overseas by airplane. The United States maintains forward military presence quite inefficiently today, generally needing four to six vessels in the fleet to maintain one on station in a place like the Mediterranean Sea or Persian Gulf. During the cold war, when a large Navy was desired in case of global war with the Soviet Union, that inefficient practice was tolerated. Today, with the Navy's size largely being determined by the forward presence mission, it should not be. Instead, crews could train on one ship in U.S. waters, fly to an overseas port (by commercial or DoD carrier), and subsequently relieve a crew that was manning another ship of the same class. Even allowing large cushions for warfighting purposes, sanctions enforcement, and other missions, this approach would allow the Navy to reduce its carrier fleet by one-third and its overall battle fleet by about 15 percent.

Heighten the distinction between weapons procurement and weapons modernization. Increasing the former is an urgent priority for a force with much aging equipment that will soon become unsafe and unreliable. Accelerating the latter, given the overwhelming dominance of U.S. military forces in relation to any possible foe today, is less time-critical and can be done more selectively. Programs like the various tactical combat aircraft now being developed have generally progressed far enough to justify some production, but they do little to redress the Pentagon's greatest likely vulnerabilities in future war: ballistic or cruise missile attack, shallow-water mines, guerrilla combat threats, and the use of weapons of mass destruction by adversaries. Modernization should focus less on major platforms and more on these vulnerabilities, as well as high-payoff, lower-cost systems such as improved communications, computers, and munitions.

Reduce nuclear forces to 3,500 warheads immediately. The nuclear-capable B-1 fleet, now envisioned for conventional missions, would remain a hedge against indefinite Russian nonratification of the START II Treaty; by restoring it to dual-capable service, the United States could return its strategic warhead count to about 5,000 if that ever seemed necessary. Given the amount of destructive power in the U.S. nuclear arsenal, and the fact that U.S. forces are in far better condition than Russia's and far more survivable against any hypothetical first strike, this position would not be risky on military or political grounds. This hedged drawdown in the nuclear arsenal would be similar in spirit to former president George Bush's 1991 unilateral steps to reduce certain types of nuclear weapons and change the deployment patterns of others. The United States should also offer to reduce the alert levels on many of its remaining weapons if Russia would reciprocate. Net savings from such nuclear cuts are likely to be small, however, since any funds freed up by force cuts may be needed to increase spending on missile defenses within a few years.

These ideas are developed further in the rest of the book. Chapter 2 completes the background section, putting U.S. defense budgets in international, historical, and fiscal context. Its broad message is that current U.S. defense spending is neither abnormally high nor unacceptably low. Given those conclusions, and the likely nature of future budget debates, DoD will probably not suffer large cuts again soon—but neither will it likely benefit from significant real spending increases after 2002. In that case, DoD will lack the money to sustain its current force and strategy much beyond 2002, so current plans will have to be modified. Chapter 2 also addresses the important budgetary questions of defense privatization and NATO expansion.

Chapters 3 through 5 lay out an alternative concept for the future U.S. military that addresses the coming budget crunch. Chapter 3 presents the core of the alternative. Specifically, it critiques the so-called two–Desert Storm warfighting approach, standard naval operations, and nuclear weapons priorities. The corresponding changes in force structure that it advocates would reduce active-duty military personnel by about another 100,000 to 1.25 million total troops. Chapter 4 examines a number of high-profile weapons programs and argues that several of the most expensive should be scaled back while others, generally less expensive, should be expanded. Among those programs where economies seem appropriate are the F-22 Raptor fighter, F/A-18 E/F Super Hornet combat aircraft, Comanche scout/light attack helicopter, and V-22 Osprey tilt-rotor transport aircraft. The DDG-51 destroyer and D5 submarine-launched missile programs should be ended outright. Resulting gross savings would total roughly $9 billion a year, although about half that amount of funding would be needed to expand purchases of certain other types of equipment, such as fast sealift ships and current-generation fighters and helicopters. Chapter 5 considers the broad issue of readiness and personnel funding and shows that these accounts will indeed continue to require the large number of dollars now projected for them. Indeed, there is a good case for a military pay raise, above and beyond the rate of inflation, that is not now in the QDR plan.

The net savings from this alternative force posture would be about $15 billion. That may not be quite enough to hold real national security (050) spending at the planned 2002 level of $250 billion, but such a force would certainly be affordable at the scheduled 2003 spending level of $260 billion.

2
THE U.S. DEFENSE SPENDING CONTEXT

Although broad measures of defense spending trends cannot resolve the question of what level of resources the United States should devote to its military, they can provide a valuable perspective that many policymakers and other citizens do not currently have. This chapter provides such perspective. It also sheds light on likely future U.S. defense spending debates in the context of efforts to balance the federal budget by 2002 and keep it balanced thereafter.

To the extent the following data and trends have a message, it is probably that one cannot resolve the matter of how much the United States should spend on defense in the future from broad spending indicators. The reader would be well advised not to heed arguments made primarily on the basis of such highly aggregated data comparisons. U.S. armed forces enjoy a remarkable degree of superiority over any other country, and most of the world's second-tier powers are U.S. allies. It is also true, however, that U.S. defense costs are now less than during the cold war, and much less as a percentage of national economic output. In addition, international and historical comparisons do not begin to wrestle with two important asymmetries in American defense planning: this country is the only one in the world that takes on a wide range of global security responsibilities, and it also believes in winning wars decisively to minimize casualties. Complaints from the left that cold war defense spending habits and levels continue, and from the right that the once-proud U.S. armed forces are being misused by irresponsible politicians and suffering a serious degradation of their near-term readiness due to lack of resources, tend to be equally unconstructive.

THE DEFENSE DRAWDOWN

Where does U.S. defense spending stand in the context of the post-1990 military drawdown and the overall federal budget? Most of the following discussion is in terms of spending or outlays—what actually flows out of the U.S. treasury each year in support of national security. It is that number, rather than the "budget authority" that gives the military the right to enter into new financial contracts each year, that figures in calculations of the deficit and is directly constrained under future plans for a balanced budget.

In fiscal year 1998 U.S. national security spending will be roughly $265 billion. That number includes not only Department of Defense (DoD) spending but also about $12 billion dedicated to nuclear weapons research and maintenance as well as environmental cleanup within the Department of Energy's budget. Defined in this way, the national security account is also known as the "050" federal budget function. In terms of constant 1998 dollars, used throughout this text unless otherwise noted, the $265 billion level is roughly $100 billion below defense outlays at the end of the cold war (see table 2-1). Viewed another way, it is roughly $50 billion below the overall cold war average.

Although press reports have often claimed that defense spending has been protected or even increased by recent budget accords, the post–cold war defense drawdown is not yet quite over. Real 050 spending is still going down each and every year, albeit slightly. It is about 10 percent less in 1998 than in the year when the Republican-led 104th Congress came to town (fiscal year 1995). Given projected inflation rates, it will decline to roughly $250 billion by 2002, the final year of the 1997 White House–Congress budget accord, before recovering some of those losses and winding up at $260 billion in 2003 (as expressed in constant 1998 dollars) according to the administration's latest long-term budget plan.

Some analysts, pointing to growth in Pentagon spending on activi-

TABLE 2-1. *National Security Spending*[a]

Billions of constant 1998 dollars

Year or period	Outlays
1997	277
1998, estimated	265
2002, projected	250
2003, projected	260
Historical perspective	
Cold war average	320
1980s peak (1989)	390
1990	375
1982–91, ten-year average	360
Bush "Base Force" plan for 1997	285
Expected requirements	
2002–10, BUR	280
2002–10, QDR	270

Sources: *Budget of the United States Government, Fiscal Year 1999: Historical Tables,* pp. 117–18; Congressional Budget Office, "An Analysis of the Administration's Future Years Defense Program for 1995 through 1999," CBO Paper (January 1995), p. 50; Office of the Under Secretary of Defense (Comptroller), *National Defense Budget Estimates for FY 1998* (Department of Defense, 1997), pp. 49, 108–11; and Michael O'Hanlon, *Defense Planning for the Late 1990s* (Brookings, 1995), pp. 38–40.

a. Includes defense-related Department of Energy activities.

ties like environmental cleanup and peacekeeping operations, argue that resources being devoted to national defense in the strict sense of the word have been cut even more than these numbers suggest. Indeed, these added costs constitute about $10 billion a year more of current national security outlays than in past decades.[1]

Cold war spending levels were, however, inflated in their own ways that had little to do with equipping, training, and routinely operating the country's main combat units. First, nuclear weapons

1. See Stephen Daggett and Keith Berner, "Items in the Department of Defense Budget That May Not Be Directly Related to Traditional Military Activities," Congressional Research Service memorandum, March 21, 1994, p. 39; Department of Defense, "FY 1998 Defense Budget Briefing," February 6, 1997; Ellen Breslin-Davidson, "Restructuring Military Medical Care," Congressional Budget Office, July 1995, p. 16; Deborah Clay-Mendez, Richard L. Fernandez, and Amy Belasco, "Trends in Selected Indicators of Military Readiness, 1980 through 1993," Congressional Budget Office, March 1994, p. 38.

expenditures were generally about $20 billion higher than today.[2] Second, the cold war average also includes actual expenditures for two regional conflicts of a type not present in the current budget (supplemental funds would be needed if the country went to war today). A conservative estimate is that the Korean War cost the United States $150 billion more than it likely would have spent on defense over fiscal years 1952–54 otherwise. That total is calculated by comparing defense spending during the war with subsequent Eisenhower administration spending levels. The Vietnam War cost the country at least $250 billion. (In this case, for sake of reaching a conservative estimate, the period just before the war is taken as the benchmark for comparison, rather than the post-Vietnam years when 050 accounts dropped precipitously.) These wars together account for at least $10 billion of the cold war annual average.[3] If anything, the end of the cold war has resulted in more hidden savings than hidden costs in DoD's budget.

A brief word on budget authority: in 1998, its $268 billion level is nearly equal to actual spending of $265 billion. But as procurement funding increases over the next few years, the gap between spending and budget authority will temporarily widen. Budget authority will remain steady in real terms even as spending declines. By 2002, budget authority will exceed spending by $15 billion, and in 2003 it will still lead spending by almost $10 billion.[4] Thereafter, once procure-

---

2. See Stephen I. Schwartz, U.S. Nuclear Weapons Cost Study Project, "Maintaining Our Nuclear Arsenal Is Expensive and Dangerous," *Washington Times*, March 26, 1997; David Mosher and Michael O'Hanlon, *The START Treaty and Beyond* (Congressional Budget Office, October 1991), p. 135; and Michael O'Hanlon, "Implementing START II," Congressional Budget Office, March 1993, p. 34.

3. Office of the Under Secretary of Defense (Comptroller), *National Defense Budget Estimates for FY 1998* (Department of Defense, 1997), pp. 108–09.

4. Specifically, only about 15 percent of a given year's budget authority for procurement is typically spent that year. Roughly 30 percent is outlayed each of the next two years, and again about 15 percent the fourth year (though actual amounts vary by type of purchase—for example, shipbuilding accounts feature slow outlays with significant spending occurring even six and seven years after funds are authorized). By contrast, 70 percent of budget authority for personnel is spent the first year, and almost all the rest the second year; for operations and maintenance accounts, the figures are roughly 50 percent the first year, 40 percent the second, and 5 percent the third; for research, development, testing, and evaluation, the respective average rates are roughly 45 percent, 40 percent, and 10 percent. See *National Defense Budget Estimates*, pp. 60–61.

ment budgets plateau, spending and budget authority will presumably converge once again.

Budget authority is the most revealing measure of year-to-year variation in the defense budget because it is what the Congress directly controls. In that light, before leaving the subject it is worth noting that since taking power in 1995, the Republican Congress added a total of roughly $21 billion in budget authority to Pentagon coffers relative to what the administration had requested each year—some $7 billion in the fiscal year 1996 budget, $11 billion for 1997, and $3 billion for 1999.[5] In general, however, the differences between budget authority and spending are modest enough that an examination of broad trends in defense resources can focus on the latter, as is done in the rest of this chapter.

Viewed as a percentage of gross domestic product (GDP) rather than in real dollar terms, defense spending is declining even faster (see figure 2-1). It is hard to view this trend as dangerous; there is little if any reason to think that U.S. defense spending should be linked to the size of the nation's overall economic output. It is nevertheless unpersuasive to argue that U.S. defense spending can be considered a significant strain on the economy. Only 3.2 percent of GDP in 1998, it is scheduled to decline further to just 2.8 percent of GDP by 2003, in contrast to levels of about 6 percent in the 1970s and 1980s and 8 to 10 percent in the 1950s and 1960s.[6]

DEFENSE AND THE FEDERAL BUDGET

Making a thumbnail sketch of the last forty years, defense spending represented almost half the federal budget throughout the 1960s, declined throughout the 1970s to just under a quarter of federal spending, was restored to slightly more than a quarter in the 1980s, and has been declining ever since. As of 1998 it represents 16 percent of the federal budget, a figure that will drop to 15 percent by 2002 (see table 2-2).

5. Department of Defense, "FY 1999 Defense Budget Briefing," February 2, 1998; *Budget of the United States Government, Fiscal Year 1998*, p. 217; and *Budget of the United States Government, Fiscal Year 1999*, p. 261.

6. *Budget of the United States Government, Fiscal Year 1999: Historical Tables*, p. 118.

FIGURE 2-1. *U.S. Defense Spending Relative to Gross Domestic Product, Fiscal Years 1947–2003*[a]

Percent of GDP

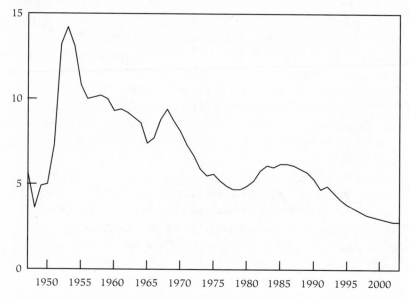

Sources: Department of Defense, Office of the Under Secretary of Defense (Comptroller), *National Defense Budget Estimates for FY 1998* (March 1997), pp. 172–73; and *Budget of the United States Government, Fiscal Year 1999: Historical Tables*, p. 117.

a. The 1998 spending level is 3.2 percent; the 2003 spending level will be 2.8 percent.

Other broad trends in the federal budget over the past generation include the relative rise and fall of domestic discretionary spending, the continued decline of international affairs spending, and the steady growth through the mid-1990s of federal interest payments, which are expected to taper off somewhat by 2002. These trends roughly cancel out over the 1962–2002 period. Therefore, the main budget story over those four decades is a massive shifting of federal resources from defense to entitlements. Such "mandatory" accounts—dominated by social security, medicare, and medicaid—will have risen from about 25 percent of the federal budget to more than 55 percent of the total by 2002.[7]

Domestic discretionary accounts—which fund such items as federal highways, health research, NASA, housing programs, criminal

7. *Budget of the United States Government, Fiscal Year 1999: Historical Tables*, p. 117.

TABLE 2-2. *Selected Categories of U.S. Federal Spending,*
1962–2003
Percent of total

Category	1962	1980	1990	1998	2003
Social security	13	20	20	23	24
Defense discretionary spending	49	23	24	16	15
Domestic discretionary spending	13	22	14	16	15
Net interest payments	6	9	15	15	11
Medicaid	0	2	3	6	7
Other means-tested entitlements	4	6	5	7	7
Medicare	0	6	9	13	15
International affairs	5	2	1.5	1.1	1.0

Sources: Congressional Budget Office, *Economic and Budget Outlook: Fiscal Years 1999–2008* (January 1998), pp. 70, 120; *Budget of the United States Government, Fiscal Year 1999, Historical Tables*, pp. 117, 274.

justice, and veterans' health care benefits—increased about 20 percent in real terms between 1990 and 1997. Budget authority for those accounts then increased more than 5 percent from 1997 to 1998. That was partly a counterreaction to the cuts they suffered in the 1980s. Today, this domestic budget category represents about the same share of GDP as in the Reagan years and well below that of the Nixon years. Arguably, much of the domestic discretionary account should be viewed relative to the size of the nation's GDP—or at least relative to the size of its population—since much of it responds to demands for infrastructure, education, environmental cleanup, crime control, and other priorities that are linked to the amount of human and economic activity in the country. By 2002 discretionary spending will represent a share of GDP that has not been so low since early in the Kennedy administration.[8]

U.S. DEFENSE SPENDING IN INTERNATIONAL PERSPECTIVE

In 1989 the world spent about $1,350 billion on defense, according to U.S. government estimates (in 1998 dollar terms). Of that total, the

8. *Budget of the United States Government, Fiscal Year 1999: Historical Tables*, 118; and Eric Pianin, "Budget Pact's 1st Bottom Line: A Surge in Domestic Spending," *Washington Post*, November 26, 1997, p. A1.

United States accounted for $390 billion, or 29 percent; NATO accounted for $600 billion, or just under 50 percent. Again according to U.S. government estimates, the Soviet Union spent $400 billion, or 30 percent; the Warsaw Pact countries together spent $475 billion, or 35 percent; and China was estimated to be spending about $30 billion. Most of the remainder of the world's spending came from the rest of East Asia and the Middle East.[9]

By 1995 the picture was much different, as it remains today. Total global military spending was down to about $900 billion (or, if expressed in 1995 dollars, about $800 billion, as shown in table 2-3).[10] More significantly, the bipolar structure of the cold war had been replaced by a unipolar/U.S.-centered alliance structure. The U.S. share of total global spending was up about 5 percent, and NATO's was up about 10 percent. When Japan, the Republic of Korea, and Australia are factored in, the western alliance system as a whole accounted for 66 percent of global defense spending (see table 2-3). Adding in other allies and friends of the United States, what might be more loosely called the western community accounted for 77 percent of total world military expenditure. The so-called rogue states as a group accounted for less than 2 percent of global spending; China and Russia about 13 percent between them (slightly higher by U.S. government estimates, as discussed below); and other Asian countries most of the rest.

Defense Burdens within the Western Alliance System

As noted, the United States accounts for about 33 percent of global military spending and its major formal allies for a similar amount (see tables 2-3 and 2-4). Japan, France, Germany, and the United Kingdom, plus Italy and South Korea in a second tier of spending,

9. Arms Control and Disarmament Agency, *World Military Expenditures and Arms Transfers, 1990* (1991), pp. 2, 3, 51, 58, 85.

10. Here the estimates of the International Institute for Strategic Studies are employed, since they are generally more up to date than those of the U.S. Arms Control and Disarmament Agency. The estimates differ most for China—by roughly $20 billion, ACDA's number being higher. See discussion later in this chapter, as well as Arms Control and Disarmament Agency, *World Military Expenditures and Arms Transfers, 1995* (1996), p. 42; International Institute for Strategic Studies, *The Military Balance 1995/96* (Oxford University Press, 1995), p. 176; and IISS, *Military Balance 1996/97*, p. 179.

TABLE 2-3. *Global Distribution of Military Spending, 1996*

Countries	Defense spending (billions of 1995 dollars)	Percent of global total	Cumulative percent (running total)
United States and its major security partners			
United States	265.8	33.4	33
NATO (not including the United States)	191.6	24.1	57
Major Asian allies[a]	67.2	8.4	66
Other allies[b]	30.7	3.9	70
Other friends[c]	55.7	7.0	77
Others			
Russia	69.6	8.7	85
China	34.7	4.4	90
"Rogue states"[d]	11.8	1.5	91
Remaining Asian countries	45.5	5.7	97
Remaining European countries	10.0	1.3	98
Remaining Middle Eastern countries	6.1	0.8	99
Others[e]	7.8	1.0	100
Total	796.5	100.0	100

Source: International Institute for Strategic Studies, *The Military Balance 1997/98* (Oxford University Press, 1997), pp. 293–98.

a. Japan, South Korea, and Australia.

b. New Zealand, Thailand, Philippines, and the Rio Pact countries minus Cuba (including all South American countries except Belize, Guyana, and Suriname, and also four Caribbean islands or island groups: the Bahamas, the Dominican Republic, Haiti, and Trinidad and Tobago).

c. Austria, Czech Republic, Hungary, Ireland, Poland, Sweden, Switzerland, Israel, Egypt, Jordan, Kuwait, Oman, Qatar, and Saudi Arabia.

d. Cuba, North Korea, Iran, Iraq, and Libya.

e. Principally African and Caribbean countries.

account for 75 percent of the allies' total; adding Canada, the Netherlands, Spain, and Australia to the list pushes that figure to 90 percent.

As can be seen in table 2-5, as of 1995 the United States was devoting nearly four times as much of its economic output to defense as was Japan, and nearly two-thirds again as much as NATO Europe. Even when foreign aid is considered, the United States was spending 50 percent more on international activities than was NATO Europe, and roughly three times as much as Japan, as a percentage of GDP.

TABLE 2-4. *Defense Spending by NATO and Major Formal U.S. Allies, 1996*

Country	Defense expenditures (billions of 1995 dollars)	Percent of GDP	Size of armed forces (thousands)
NATO			
United States	265.8	3.6	1,483.3
France	46.2	3.1	398.9
Germany	38.4	1.7	358.4
United Kingdom	32.7	3.0	226.0
Italy	23.3	2.2	325.2
Canada	8.4	1.5	70.5
Netherlands	7.9	2.1	63.1
Spain	8.4	1.5	206.8
Turkey	6.8	3.9	525.0
Greece	5.4	4.8	168.3
Belgium	4.2	1.6	46.3
Norway	3.7	2.4	30.0
Denmark	2.9	1.7	32.9
Portugal	2.8	2.8	54.2
Luxembourg	0.1	0.7	0.8
Iceland	. . .	. . .	. . .
Total, non-U.S. NATO	191.2	2.2	2,506.4
Total, NATO	457.0	2.3	3,989.7
Other major formal U.S. allies			
Japan	43.6	1.0	235.5
South Korea	15.1	3.3	660.0
Australia	8.4	2.2	57.8
Total, other major U.S. allies	67.1	2.2	953.3
Grand total	524.1	2.3	4,943.0

Source: IIIS, *The Military Balance 1997/98*, pp. 293, 295.

China and Russia

Because of the difficulty of comparing spending levels between fundamentally different types of economies, and because of the importance of understanding defense spending levels in China and Russia, a fuller discussion is in order for these matters.

Officially, Russia budgeted about $18 billion for its military in 1996, based on the ruble-dollar exchange rate. The equivalent of another $5 billion was provided for internal security, defense research and development, arms control and demilitarization, and other

TABLE 2-5. *Overall Foreign Policy Spending as Percent of GDP, 1996*

Type of spending	NATO Europe (excluding Greece and Turkey)	Canada, Australia, New Zealand, Sweden, Switzerland, and Austria	Japan	United States
Official development assistance	0.37	0.37	0.20	0.12
Other aid-related activities, including peacekeeping (approximate)	0.075	0.025	0.01	0.1
Defense spending	2.2	1.9	1.0	3.7
Total	2.6	2.3	1.2	3.9

Sources: Organization for Economic Cooperation and Development, "Aid and Other Financial Flows in 1996," June 18, 1997; and International Institute for Strategic Studies, *The Military Balance 1996/97* (Oxford University Press, 1996), p. 40.

related costs that standard NATO definitions would consider to be national security spending. So the total was about $23 billion.

China's official 1996 budget was $8.4 billion, but that figure only includes a modest fraction of what would normally be considered total military spending. It appears to exclude the budgeted costs of most military research and development and weapons acquisition programs, salaries for civilian employees of the armed forces, military construction expenses, and certain revenues that the armed forces are allowed to raise and spend on their own. An adjusted total would be about two and a half times as high, or roughly $20 billion.[11]

Even having made these corrections, defense budget estimates for China and Russia based on official exchange rates are too low. The main reason is that they do not reflect real spending power within the countries at issue. The resulting systematic understatement of their budgets can be corrected through use of purchasing-power parity (PPP) comparisons. Those comparisons account for the simple fact that countries like Russia and China "get a discount" relative to what would be required to support the same number of troops in the West because many goods that any military and its soldiers need cost less

11. See IISS, *Military Balance 1995/96*, pp. 176, 272; and IISS, *Military Balance 1996/97*, pp. 107–13, 179.

in poorer societies—even more so than implied by the official exchange rate. If one is going to use defense spending as even a rough guide to evaluating military balances and assessing threats, PPP corrections should be used. It makes little sense to claim a military advantage because an American soldier's quarter-pounder costs more in dollar terms than a Chinese soldier's lunch or because a house in Santa Barbara costs more than one in Vladivostok.

Unfortunately, PPP measures are difficult to compute. The main challenge is comparing the quality of goods between countries. Getting enough detailed data to do a careful comparison is difficult and often impossible. As such, even if the relative prices of goods are known in different countries, their relative value often is not. As a consequence, reputable PPP estimates often vary by a factor of two; they can even vary by a factor of five or more.

These facts help explain why a recent RAND study could argue, albeit controversially and not particularly persuasively to most, that China spends $150 billion a year on its military.[12] They also explain why the U.S. government estimates China's 1995 defense spending at slightly less than $70 billion, as expressed in 1998 dollars—nearly twice the $35 billion figure estimated by the International Institute for Strategic Studies (IISS) for that year or the $40 billion estimated by IISS for 1996. (The U.S. government estimates Chinese military spending as 2.3 percent of GDP, roughly the worldwide median by that metric.)[13]

The key point is that any comparison of U.S. to Russian and Chinese defense spending levels is highly inexact. For example, even ignoring the more extreme possible estimates, the IISS's 1996 U.S.-Russian defense spending ratio of 3.5:1 might be more fairly expressed as somewhere between 2.5:1 and 5:1.[14] The ratio of U.S. to

12. See Charles Wolf Jr. and others, *Long-Term Economic and Military Trends, 1994–2015* (Santa Monica, Calif.: RAND Corporation, 1995), p. 15.

13. The U.S. official estimate is for 1995 and is expressed in terms of constant 1998 dollars. See U.S. Arms Control and Disarmament Agency, *World Military Expenditures and Arms Transfers, 1996* (1997), pp. 36, 39; and IISS, *Military Balance 1997/98*, p. 176.

14. The possibility that Russia's adjusted military budget could be even larger than $80 billion might be worrisome were it not for its general downward direction, the poor state of Russian equipment, the decline of morale among troops, and the collapse of the procurement budget. See, for example, Clifford Gaddy, *The Price of the Past: Russia's Struggle with the Legacy of a Militarized Economy* (Brookings, 1996), p. 171.

Chinese spending might be best thought of as somewhere between 3:1 and 10:1. On the other hand, if a consistent methodology is used, PPP measures can do a reasonably good job of tracking trends in a country's military spending from year to year. In any case, they give a better relative sense of different countries' defense resources than do exchange rate–based budget comparisons.

Regardless of the precise numbers, the main message is the same: global defense spending remains heavily weighted in the favor of the West in general and the United States in particular. Even at the higher plausible estimates of their defense spending, China and Russia together account for less than one-fifth of global military expenditure—whereas the United States spends about one-third of the total and its friends and allies at least another third.

Budget Pressures and Future Requirements

Can the Quadrennial Defense Review (QDR) force be maintained, given officially planned budgets over the next five years as well as the resources likely to be available thereafter? While all one can really do now is speculate, the most probable answer appears to be no.

In the short term, defense spending appears to be safe from cuts beyond those already scheduled. It is possible to make arguments for why those resources could go up or down over the next five years, but on balance the most likely course would seem to be that the spending trajectory recently codified in the Congress's and president's budget deal will be followed. Beyond 2002 the picture is fuzzier. The most likely scenario, however, is that spending will be inadequate to fund the QDR force. Real defense spending increases would probably need to reach $270 billion, which they are not scheduled to do even after a $10 billion spending increase in 2003 (see table 2-1). It is possible, as discussed further below, that less money might prove adequate to fund the QDR force, but it is more likely that a steady-state sum in excess of $270 billion will be needed.

The Near-Term Budget Environment

Leaving aside strategy debates over whether the QDR force has the right size and composition for the demands of the post–cold war world, the amount of money allocated to defense in the 1997 budget deal should be about enough to keep DoD fit, safe, and ready through 2002. Military compensation, although not quite as good as in the Reagan years, will at least hold its own relative to the Bush administration standards of the early 1990s. Readiness funding will remain robust, thanks to a major adjustment in plans that was made during the preparation of the 1999 budget request.

There is a chance that pressure to boost defense spending at least slightly above planned levels will arise. For example, House Speaker Newt Gingrich—despite having described himself as a "cheap hawk" in 1995 and despite being a principal player in the balanced-budget agreement—subsequently advocated relaxing its constraints on defense spending.[15] The most likely source of any such political pressure in the future is likely to be the continued dearth of funds available for equipment purchases under existing administration plans.

It is also possible that defense spending could be cut again between now and 2003. For example, an economic downturn, perhaps precipitated by the slump in Asia, could increase pressure to cut spending in order to stay in fiscal balance.[16] Similarly, specific assumptions and goals of the 1997 balanced-budget act may prove optimistic and keep the federal government in the red longer than expected, forcing budget balancers to look around for additional sources of savings. Notable on the list of uncertainties are the planned cuts in medicare, anticipated revenues from sales of rights to the electromagnetic spectrum, and changes in revenues from the new tax policies put in place.[17] In 1997 a majority of House Democrats signed a letter advocating deeper defense cuts in the course of the balanced-budget negotiations; similar efforts may be made again.

But the most likely political outcome would seem to be compliance with the budget deal for the next few years. For example, the above-

15. Newt Gingrich, "Defense Matters," *National Review*, February 9, 1998, p. 44; and Philip Finnegan, "U.S. Defense Spending Hike Support Wanes," *Defense News*, January 26–February 1, 1998, p. 3.

16. Patrice Hill, "Surpluses Not Assured, Fed Chief Cautions Hill," *Washington Times*, January 30, 1998, p. A1.

17. Robert D. Reischauer, "Those Surpluses: Proceed with Caution," *Washington Post*, September 21, 1997, p. C9.

noted Democrats had no Republican cosignatories and represented but one-fourth of the House. Also, President Clinton actually pushed for higher defense spending in the 2000–02 period than the Republican Congress's own balanced-budget proposal would have allowed, and the budget deal ultimately reflected that higher level (as well as Congress's preference for higher defense spending than Clinton wished in 1998 and 1999).[18]

In addition, the continually improving economic and budget outlook has reduced the need for deep cuts in different parts of the budget to achieve and sustain fiscal balance. That in turn should reduce policymakers' inclinations to raid defense accounts in order to provide relief to various domestic programs. For example, both the president's and Congress's 1996 proposals to reduce real annual funding for domestic discretionary programs by more than 20 percent by 2002—described as "the unfulfillable promise" by former Congressional Budget Office director Robert Reischauer—have been dropped. President Clinton's 1999 budget request would cut this spending only about 5 percent between 1998 and 2003.[19]

Some recent polls suggest that most Americans would support a further modest cut in defense. But as noted, such a cut is already scheduled under the balanced-budget deal (though it is not clear that most Americans realize that fact yet). For example, the University of Maryland's Program on International Policy Attitudes suggested that Americans would on average support about a 10 percent cut in defense spending relative to the 1995 level. But the 1998 spending level is already 10 percent less than 1995 outlays, and by 2003 the level is scheduled to drop another 2 percent.[20]

18. Sheila Foote, "Majority of House Democrats Urge Defense Spending Cuts," *Defense Daily*, April 28, 1997, p. 163.

19. Robert D. Reischauer, "The Unfulfillable Promise: Cutting Nondefense Discretionary Spending," in Robert D. Reischauer, ed., *Setting National Priorities: Budget Choices for the Next Century* (Brookings, 1997), p. 145; press briefing by the president's budget team, The White House, May 2, 1997; and *Budget of the United States Government, Fiscal Year 1999, Historical Tables*, p. 118.

20. Steven Kull, I. M. Destler, and Clay Ramsay, *The Foreign Policy Gap: How Policymakers Misread the Public* (College Park, Md.: University of Maryland Program on International Policy Attitudes, 1997), p. 121.

Medium-Term Budget Prospects

The authors of the QDR built their plan on the assumption that real defense resources for budget 050, slated to decline to about $250 billion by 2002, will not rise much above that real level thereafter.[21] In that judgment, they were wise. Unfortunately, the plan they developed appears likely to cost about $270 billion a year in the next decade and beyond.

The need to boost 050 spending arises principally from the fact that procurement accounts will need to rise as systems purchased during the 1970s and the 1980s wear out. The procurement account has historically been about one-third the size of the rest of the defense budget in aggregate; at present it is only one-fifth as large as the others (see table 2-6). That is too low to be sustainable—even if a more frugal approach is taken to buying new weaponry.

It is for this basic reason that the Joint Chiefs of Staff have advocated a $60 billion procurement spending level for several years. The administration now claims the goal will be achieved in the early years of the next decade. Because of the steady push of inflation, however, $60 billion three to five years from now will be worth considerably less than it would have been when the goal was first articulated in the mid-1990s—some 15 to 20 percent less.[22]

Moreover, restoring the procurement account to 25 percent of total defense spending would require at least $65 billion for a force of the anticipated size and cost, as indicated in figure 2-2, or $10 billion more than now planned for 2003. Indeed, since weapons costs typically increase faster than manpower expenses, future procurement costs may exceed 25 percent of the total DoD budget and necessitate an even larger boost in funding. Fortunately, after 2003 the Pentagon should begin to realize savings from its additional rounds of base closures scheduled for 2001 and 2005 that will eventually reach $3 billion a year. Some other efficiences should be achieved as well. So a net increase in national security spending of $10 billion may suffice. But that still leaves an average annual price tag of about $270 billion for the 050 account.

21. William S. Cohen, *Report of the Quadrennial Defense Review* (Department of Defense, May 1997), p. 59.

22. "Military Splits with Clinton on Arms," *Washington Times*, March 14, 1996, p. 4.

TABLE 2-6. *Department of Defense Budget Authority, 1998–99*[a]
Billions of dollars

Category	1998 (actual)	1999 (requested)
By title		
Military personnel	69.7	70.8
Operations and maintenance[b]	94.4	94.8
Procurement	44.8	48.7
Research, development, testing, and evaluation	36.6	36.1
Military construction	5.1	4.3
Family housing	3.8	3.5
Other[c]	0.5	–0.8
Total	254.9	257.3
By major service		
Army	60.5	63.8
Navy[d]	80.9	81.3
Air Force	74.4	76.7
DoD-wide	39.0	35.4
Total	254.9	257.3

Source: Office of the Assistant Secretary of Defense (Public Affairs), "DoD Budget for FY 1999," February 2, 1998, pp. 6–7; and R. Jeffrey Smith, "Espionage Budget Totaled $26.6 Billion," *Washington Post*, October 16, 1997, p. 1.

a. Figures may not add to totals due to rounding. Department of Energy 050 spending is not included here. DoD released, for the first time publicly, information that intelligence spending was $26.6 billion in 1997. A similar amount is undoubtedly found within the above defense budget totals (in various of the above-mentioned accounts).

b. The operations and maintenance account funds civilian pay.

c. Includes receipts.

d. Spending of about $10 billion for the Marines is included within the Navy total.

Recent CBO studies provide a more detailed and rigorous confirmation of this projection. In 1995 the CBO estimated that the average annual cost of the Bottom-Up Review next decade would be roughly $265 billion to $285 billion (when expressed in 1998 dollars), the uncertainty being due principally to questions about what planned weapons systems would wind up costing. Although the Pentagon insists that using more commercial goods and implementing additional acquisition reforms will largely eliminate the cost growth that has traditionally affected its weapon programs, there is good reason to doubt that optimism. Grounds for skepticism have been recently reinforced by the Pentagon Program Analysis and Evaluation Office's recent estimate that the F-22 "Raptor" aircraft would cost about 25 percent more

FIGURE 2-2. *Department of Defense Procurement, 1948–2003*

Billions of FY 1998 dollars of outlays

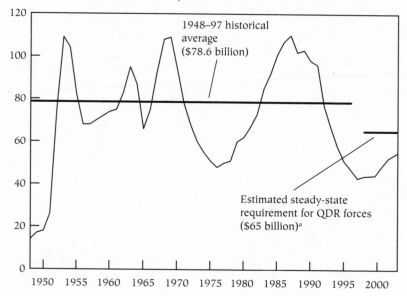

Sources: Department of Defense, *Report of the Secretary of Defense to the President and Congress* (April 1997), pp. 108–11; and Department of Defense, "FY 1999 Defense Budget: Backup Charts and Briefing," February 1998.

a. The $65 billion estimated requirement is based on Congressional Budget Office, "An Analysis of the Administration's Future Years Defense Program for 1995 through 1999," CBO Paper (January 1995), p. 50; Michael O'Hanlon, *Defense Planning for the Late 1990s* (Brookings, 1995), pp. 38–40; and William S. Cohen, *Report of the Quadrennial Defense Review* (May 1997), pp. 60–63.

than the Air Force assumes.[23] As shown earlier, the QDR force should be about $10 billion a year cheaper than the BUR in the steady state, implying a steady-state cost of $255 billion to $275 billion. The draft results of a more recent CBO study suggest that national security spending under the QDR force might be even greater. Once Department of Energy nuclear weapons costs are factored in, the CBO

23. Weapons systems often go up in cost by 50 percent relative to initial expectations. Rachel Schmidt, "An Analysis of the Administration's Future Years Defense Program for 1995 through 1999," Congressional Budget Office, January 1995, pp. 10, 50. On the F-22, see Bryan Bender and Tom Breen, "Cohen Says He Will Trim F-22 Program If Costs Increase," *Defense Daily*, April 4, 1997, p. 25.

FIGURE 2-3. *Department of Defense Budget Authority, 1974–2014*[a]

Billions of 1998 dollars

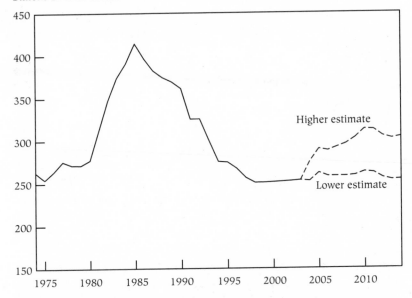

Sources: Office of the Under Secretary of Defense (Comptroller), *National Defense Budget Estimates for FY 1998* (March 1997), pp. 98–99; and "CBO Finds Potential $55 Billion or Higher Defense Budget Shortfall," *Inside the Pentagon*, November 6, 1997, p. 1.
a. Projections based on CBO analysis of Quadrennial Defense Review. Figures do not include the cost of the Department of Energy's nuclear weapons activities, which have typically been about $10 billion a year since the mid-1980s and generally ranged from $5 billion to $10 billion a year in earlier periods dating back to 1950.

estimates that 050 resources might need to total $260 billion to $300 billion from 2004 until 2015, as shown in figure 2-3 (the graph is for Pentagon budget authority, but, as noted, outlays and budget authority generally converge over time).[24] The midpoint estimate of these various CBO estimates is somewhat more than $270 billion.

Is it realistic to think that a fiscally constrained democracy will provide that level of resources after more than a decade of defense spending cuts? Perhaps. Discretionary spending accounts cannot forever be

24. See Elaine M. Grossman, "CBO Finds Potential $55 Billion or Higher Defense Budget Shortfall," *Inside the Pentagon*, vol. 13, November 6, 1997, pp. 1–14.

TABLE 2-7. *Major Federal Spending Categories as Percent of GDP*[a]

Spending category	1997	Projected		
		2010	2020	2030
Social security	5.0	5.0	6.0	6.0
Medicare	2.5	4.0	6.0	7.0
Medicaid	1.3	2.0	2.0	3.0
Other mandatory spending	3.2	3.0	3.0	3.0
Defense (2002 real-dollar level)	3.4	2.5	2.2	2.0
Other discretionary programs (2002 real-dollar level)	3.6	2.7	2.4	2.2
Total	19.0	19.0	22.0	23.0

Sources: Congressional Budget Office, *Long-Term Budgetary Pressures and Policy Options* (March 1997), p. xv; and Congressional Budget Office, *Economic and Budget Outlook: Fiscal Years 1998–2007* (January 1997), pp. 29, 36.

a. Most figures are rounded to the nearest percent, except some of those for 1997 and for discretionary spending. Net interest is not included here; all other budget categories are accounted for in the total.

expected to fund growth in federal entitlements. But the widespread projections for continued growth in entitlement spending next century also imply that all parts of the federal budget will remain under significant pressure (see table 2-7). Reforms in entitlement programs will not be made easily. They will probably be put in place only incrementally, meaning that an overall climate of fiscal austerity will endure.

Even though the most recent forecasts of the Congressional Budget Office project an elimination of the federal deficit by 2001 and annual surpluses exceeding $100 billion after 2005, a recession could wipe out surpluses of that size (the administration projects larger surpluses, but they would not exceed $150 billion themselves until 2006).[25] Additionally, lawmakers will be tempted to use any surpluses for tax breaks, new domestic initiatives, or efforts to reduce the federal debt while social security surpluses remain large. Without a new acute threat to U.S. security, most will probably not consider the Pentagon the most qualified recipient of the nation's good economic fortunes. In such an environment, discretionary accounts—and in

25. Congressional Budget Office, *The Economic and Budget Outlook: Fiscal Years 1999–2008* (January 1998), pp. xviii.

particular defense spending, which as argued above has little natural claim to a given share of the nation's GDP—will probably do well to grow with inflation.

WHY PRIVATIZATION AND ACQUISITION REFORM WON'T SAVE THE DAY

Are there any ways to redress likely budget shortfalls without making additional painful cuts in the already busy and sometimes over-worked U.S. armed forces? Some analysts are currently arguing that, by allowing the private sector to carry out many defense support activities like logistics, health care, and equipment maintenance, planned budgets might be able to sustain the planned force.

According to the QDR, defense support activities may now involve about 60 percent of all DoD manpower, meaning that significant improvements in how they are carried out should indeed be able to yield significant sums to sustain and modernize combat units.[26] Annual savings estimates as high as $30 billion have been bandied about, even from sources as reputable as one of the secretary of defense's advisory bodies known as the Defense Science Board in a November 1996 report.[27] But they are almost surely too optimistic, particularly for the near term. Often, such analyses simply define huge areas of defense activity as support, tally the number of employees associated with each area, and apply a single cost-savings factor based on private sector or previous DoD experience to estimate theoretically realizable savings. This approach is useful for calling attention to the potential of privatization but should never be confused with a road map for how to implement it.

Somewhat more realistically, perhaps, an August 1996 Defense Science Board report on outsourcing and privatization estimated that $7 billion to $12 billion in annual savings might be achieved by the year 2002. Its specific recommendations were less ambitious than that, however. They suggested privatization of three agencies—the

---

26. Cohen, *Report of the Quadrennial Defense Review*, p. 53.

27. Defense Science Board 1996 Summer Study, *Achieving an Innovative Support Structure for 21st Century Military Superiority* (Department of Defense, November 1996), p. ES-2.

Defense Commissary Agency, Defense Information Systems Agency, and Defense Finance and Accounting Service—with a combined current payroll of less than $4 billion a year.[28]

It is probably most accurate to think of defense reform as an ongoing, difficult, and rather tedious process than as a magic new approach. Some savings from it have already been realized and internalized into existing budget plans. For example, according to DoD, allowing more competition between private and public sectors based on the so-called OMB Circular A-76 process yielded annual savings of about $1.5 billion in 1996.[29] Such savings should continue to be achieved but will not provide large new sources of funds above and beyond what is already being anticipated.

The QDR report treats defense management reform, acquisition reform, and privatization more carefully than the Defense Science Board studies had done. It documents about a dozen specific changes that the Pentagon now intends to implement. The subsequent November 1997 Defense Reform Initiative Report continues in this vein, making thorough and convincing arguments for efficiencies like moving to a paperless defense acquisition process, increasing servicemembers' incentives to move themselves between assignments, reducing intermediate warehouses and middlemen for many routine supplies, and expanding use of special credit cards to save money on small defense purchases. Only a few hundred million dollars in annual spending is typically to be realized by these reforms.[30] Acquisition reforms that have been put in place already are saving money, but again in modest amounts—less than $500 million annually.[31]

Determining how to achieve such unglamorous and piecemeal savings is the essence of improving defense management. Over time the savings can add up, although they are unlikely to appear as quickly or

28. Defense Science Board, *Report of the Defense Science Board Task Force on Outsourcing and Privatization* (Department of Defense, August 1996), pp. 1, 6A, 11.

29. Ibid., p. 32.

30. Cohen, *Report of the Quadrennial Defense Review*, pp. 54–57; and William S. Cohen, *Defense Reform Initiative Report* (Department of Defense, November 1997), pp. 1, 7, 9, 11.

31. See Dov S. Zakheim, "Tough Choices: Toward a True Strategic Review," *The National Interest* (Spring 1997), p. 40; and Office of the Under Secretary of Defense for Acquisition and Technology, *Milspec Reform: Results of the First Two Years* (Department of Defense, 1996).

reach anywhere near the magnitude often advertised by proponents of privatization as well as acquisition reform and other defense efficiencies. The sum total of all these initiatives is expected to lead to a cut of more than 100,000 full-time DoD employees, accounting for most of the QDR's total personnel reductions and ultimately yielding some $5 billion in annual savings. The Pentagon will do well simply to achieve these already banked-upon savings over the next decade or so.

Again, two lessons emerge. First, large savings from privatization and other defense reforms can probably be realized, but they cannot be achieved without a careful game plan. Moreover, vague proposals for achieving them are likely to meet resistance or indifference from the DoD managers and officers charged with carrying them out, as RAND analysts Carl Dahlman and C. Robert Roll have recently argued in a thoughtful study.[32] Second, many savings are already being realized and factored into DoD's future budget baseline—meaning that they are no longer available to redress any shortfall that may still exist in that baseline.

WHY NATO EXPANSION WON'T BUST THE BANK

The merits, demerits, and best possible means of effecting NATO expansion have been debated at length and will not be revisited here in depth. My view is that the idea to rapidly expand NATO was undesirable because it risks worsening the chief security problem in Europe—the dubious safety of Russian nuclear weapons—to address much less acute concerns about near-term stability in central Europe. At this point in the process, more harm might be done to the western alliance and credibility of U.S. leadership by failing to admit the Czech Republic, Hungary, and Poland into NATO than would result to NATO-Russia relations by following through. That logic suggests, however, that a long pause should follow this round of expansion—and that the Baltic states and other non-Russian former Soviet republics should certainly not be admitted in the near future (prefer-

32. Carl J. Dahlman and C. Robert Roll, "Trading Butter for Guns: Managing Infrastructure Reductions," in Zalmay M. Khalilzad and David A. Ochmanek, *Strategy and Defense Planning for the 21st Century* (Santa Monica, Calif.: RAND Corporation, 1997), esp. pp. 294–306.

ably not until Russia itself has been offered clear conditions under which it could someday gain membership).

However policymakers choose to view these strategic issues, cost considerations should not be an important factor in their thinking, particularly for this stage of expansion. The estimates vary considerably, with the U.S. government estimating a total thirteen-year cost for all countries at $27 billion to $35 billion, the Congressional Budget Office estimating the expense at $60 billion to $125 billion, and NATO more recently coming up with the far cheaper price tag of about $1.5 billion over a ten-year period. But those are total costs to the entire alliance over a period of at least a decade. Even at the high end, U.S. annual costs would not be much more than $1 billion, or considerably less than one-half of 1 percent of the defense budget, and at the low end they could be just $50 million.

Consider first the U.S. official estimate, developed in a report to Congress by the U.S. delegation to NATO. It divides costs into three relatively equal categories: those for the new members to rebuild their forces and make them interoperable with NATO's current militaries; those for the existing European NATO members to make their forces more deployable for possible conflict to the east; and those to be shared by the alliance as a whole, largely for infrastructural upgrades and integrated communications capabilities. The U.S. costs would represent some 20 percent of this latter category, or around $2 billion between 1999 and 2009. Added to that would be whatever foreign assistance the United States provided to new member states to help them with their transition costs. Annual U.S. expenses would most likely be $200 million to $250 million. They would exceed $300 million only if the United States provided significant amounts of aid to the new members as they adjusted forces and built facilities to work with other NATO countries. Existing NATO members apart from the United States might spend up to $15 billion over the period. New members would incur costs of roughly that same amount, largely for new artillery, upgrades to armor, refurbished ammunition stocks and storage facilities, surface-to-air missiles, a squadron of western combat aircraft, and their share of infrastructure and communications gear.[33] However, DoD's methodology does not

33. U.S. Delegation at NATO, "Report to the Congress on the Enlargement of the North Atlantic Treaty Organization: Rationale, Benefits, Costs, and Implications," February 26, 1997.

appear to be particularly rigorous, so these figures should not be taken as the final word.[34]

A less expensive option, like NATO's, would focus just on making militaries' headquarters and air defense systems interoperable. A modest amount of funds for joint exercises would help give the expanded alliance some semblance of cohesiveness. Linked rail and road systems would at least provide the theoretical capacity for reinforcement in the event of war. But with such an approach, other upgrades would be sacrificed. The ground units of different countries would not be made more interoperable. Infrastructure for supporting large allied forces on the territories of new member states would not be procured. NATO standards for things like training hours could be viewed more as long-term goals than as immediate requirements.[35] This is a dangerous game to play for the military integrity of the alliance—but in the current threat environment it is not particularly dangerous for the actual security of the countries at issue and probably safer than placing excessive budgetary demands upon new democracies in Central Europe.

Even if such a modest investment might not jeopardize Polish, Hungarian, or Czech security, it would risk making a mockery of NATO military requirements and setting an unfortunate precedent.[36] Thus a compromise approach might be a more reasonable lower-cost option than NATO's current proposal. A more robust plan that focuses on these same limited categories of military capability might cost as much as $20 billion.[37] The U.S. share might then be $200 million to $400 million a year over ten to fifteen years.

The higher range of the CBO estimate includes about $20 billion in U.S. costs, $55 billion in expenses for the existing NATO allies, and $50 billion for new members. The majority of the costs for existing

34. For a critique of the U.S. estimate, see Ivan Eland, "The High Cost of NATO Expansion: Clearing the Administration's Smoke Screen," CATO Institute Policy Analysis 286, Washington, October 29, 1997, pp. 1–4. Although Eland is the primary author of the CBO study on this topic as well, his critique about the precision with which the administration's numbers were calculated is carefully documented and rather convincing.

35. See Brooks Tigner, "NATO Papers Belie Modest Expansion Cost," *Defense News*, December 8–14, 1997, p. 1.

36. Ibid.

37. Eland, "High Cost of NATO Expansion," p. 32.

members are due to the requirements of making their forces more fit for power projection—an objective that makes sense for missions like those in the Persian Gulf even if not for scenarios in central and eastern Europe (as discussed more in chapter 3). These are costs that, in my view, the major European NATO states in particular should be prepared to pay. But they are better viewed as the costs of NATO transformation for out-of-area missions than costs of NATO expansion. Other costs for current members, including the United States, are driven by the need to upgrade infrastructure and integrated communications. The CBO also estimates that new members would need a good deal more additional equipment than calculated by the U.S. delegation to NATO. For example, even under its more modest and less expensive option, the CBO estimates that the cost of fighter aircraft alone for new members would equal roughly $10 billion. New tank and antitank expenses add two-thirds as much again. The CBO also assumes high expenses for exercise facilities in the new member states where other countries' forces would also train.[38] The CBO's analysis is thorough and, in the end, persuasive if one thinks in terms of developing significant new military capabilities. But this is a vision of an expanded alliance that assumes more of a threat than really exists. Even if adopted, it would lead to added U.S. defense costs of no more than $1.5 billion a year.

CONCLUSION

The United States enjoys a favorable security position at the end of the twentieth century, though not one without its own challenges and potential problems on the horizon. The post–cold war defense drawdown, now nearly over, has been accomplished without damaging the force or harming the country's remaining security interests as past military downsizing efforts have generally done. Defense spending has been reduced quite considerably since 1990 but remains sufficient to leave the United States the world's only global military power and permit it to deter conflict and instability in several key overseas theaters at once.

38. Ivan Eland, "The Costs of Expanding the NATO Alliance," CBO Paper, Congressional Budget Office, March 1996, pp. 26–39.

At the same time, continued growth in entitlement spending, combined with likely demands to keep domestic discretionary programs at reasonably solid levels and juxtaposed with Americans' predictable aversion to tax increases, will constrain defense spending into the indefinite future. They will probably preclude the types of real defense spending increases that would be needed to support the QDR force and defense plan thereafter. Some difficult choices and creative policies will be required if American domestic and overseas interests and values are to be protected effectively in the future.

3

A NEW STRATEGY AND FORCE POSTURE

T he next three chapters of this study develop an alternative approach to warfighting strategy, global presence, weapons modernization, and military readiness for the U.S. armed forces. The central purpose is to find a way of keeping the Department of Defense (DoD) capable of executing its most important missions at or near the real spending level now scheduled for 2002.

In 2002, national security spending will be $250 billion under the recent budget deal negotiated by President Clinton and Congress (expressed in constant 1998 dollars). At projected rates of inflation, that 2002 spending figure amounts to a reduction of 6 percent from the 1998 level of $265 billion. Spending may rise to $260 billion in 2003, if President Clinton's 1999 budget request is a good guide, although such a peacetime increase in real defense outlays would be the first since 1989 and remains an uncertain prospect politically. In addition, should inflation pressures worsen, the number of dollars now planned for 2003 could be worth less in real terms even if they are appropriated by Congress. The bottom line is that the Quadrennial Defense Review (QDR) and its associated $270 billion annual price tag for the 050 account appear more expensive than the country will probably be willing to afford.

Tough decisions and innovative new policy ideas are needed in this environment. Accordingly, this chapter proposes three major changes in the way DoD conducts several of its most important missions: preparing for major theater war, conducting routine overseas deployments of naval forces, and deterring nuclear war as well as other uses of weapons of mass destruction. The nuclear options presented here call for fairly bold U.S. efforts to break the logjam on U.S.-Russian arms control. They could be important on security grounds, but are unlikely to be a major money saver, particularly in the near term. The other two changes together with proposals in chapter 4 would permit reductions in annual spending of at least $15 billion, largely in the areas of force structure and weapons modernization accounts (see table 3-1).

In brief, the main money-saving ideas developed in this chapter are as follows. The Pentagon's two-major-war strategy, envisioning two overlapping conflicts that could each require U.S. forces approaching the scale of those deployed for Operation Desert Storm, provides more insurance than the country needs. While crises could threaten U.S. interests in two places at once, most likely in the Persian Gulf and Korea, it is unlikely that both would erupt into large-scale conflict. Even more unlikely is the prospect that we would lose the initial battles in both places and need to repeat Desert Storm–like buildups to evict aggressors from allied territory. Given the capabilities of modern airpower, our improved vigilance and war preparations in both southwest and northeast Asia, the strength of the South Korean armed forces, and the atrophying Iraqi and North Korean threats, a smaller U.S. military and somewhat less demanding strategy would be adequate.

Instead of the two-major-war or "two–Desert Storm strategy," the country should adopt something more like a "Desert Storm plus Desert Shield plus Bosnia" posture. Such a force would be sufficient to conduct a single major offensive operation, a major defensive or holding operation that also allowed aggressive air attacks against an enemy's forces and other national assets, and a peacekeeping mission all at once. It would permit troop cuts of another 10 percent—resulting in a U.S. military of about 1.25 million active-duty troops in contrast to the QDR goal of 1.36 million.

In addition to the revision in warfighting strategy, that reduction in the size of the military would be brought about by changes in how the Navy conducts forward presence overseas. Although sea-based presence and warfighting capabilities remain important, particularly

TABLE 3-1. *Alternative Force Posture Relative to QDR*[a]

Category	Administration QDR plan	Suggested alternative	Average annual savings, 1999–2010 (billions of 1998 dollars)
		Suggested cuts	
Procurement			
Air Force F-22 "Raptor"	339	150	2.0
Navy F/A-18 E/F "Super Hornet"	550	300	1.5
Navy DDG-51 Destroyer	57	42	1.0
Army Comanche Helicopter	1,292	630	1.5
Army Longbow Apache upgrade	800	720	0.1
Marine V-22 Osprey Aircraft	360	150	1.25
Trident D-5 Missile	434	360	0.4
Navy CVN-77 Aircraft Carrier	1	0	0.4
Force structure			
Air Force wing size (aircraft)	72	65	0.75
Army active strength	480,000	440,000	1.5
Marine active strength	172,000	159,000	0.7
Marine wing equivalents	5	3	0.8
Army national guard brigades	30	21	1.0
Navy carrier battle groups/ air wings	12/11	8/7	5.0
Air Force Minuteman III missiles	500	200	1.25
Navy Trident submarines	14	9	0.6
Navy attack submarines	50	35	1.0
Navy, other ships	. . .	. . .	1.0
Other			
Department of Energy nuclear warhead research and test readiness	3 labs/ Nevada "ready"	2+ labs/ Nevada shut	0.3
Average annual savings			22.0

in the Persian Gulf and Pacific Rim regions, they could be ensured more efficiently by changing the Navy's rotation policies. Rather than sail ships from U.S. ports every six months or so to maintain a continuous presence overseas, ships should stay on forward station for longer periods while their crews are rotated by airlift (and changed while ships are docked in friendly foreign ports). This modification, though logistically challenging for the Navy, is feasible and

TABLE 3-1 *(continued)*

Category	Administration QDR plan	Suggested alternative	Average annual savings, 1999–2010 (billions of 1998 dollars)
	Suggested increases		
Procurement			
Pre-positioned brigade sets in Persian Gulf/Diego Garcia	4	6	−0.2
Navy fast sealift capacity (divisions)	3	5	−0.2
747-class Air Force airlifters	0	50	−1.25
KC-10-class Air Force refueling planes	0	50	−0.5
Air Force F-15C fighters	0	150	−0.75
Marine CH-53 lift helicopters	0	200	−0.6
Army utility helicopter initiative	. . .	. . .	−1.0
Smart munitions enhancement	. . .	. . .	−0.1
Readiness, research			
Military real pay increase	0.5%	3.0%	−1.8
Joint experimentation funds (National Defense Panel idea)	. . .	. . .	−0.5
Redeploy most Okinawa Marines	. . .	. . .	−0.3
Initiatives and other costs	. . .	. . .	−7.0
Total annual savings	. . .	. . .	15.0

Sources: Congressional Budget Office, *Reducing the Deficit: Spending and Revenue Options* (March 1997), pp. 16–86; Congressional Budget Office, "Implications of Additional Reductions in Defense Spending," October 1991, p. 16; Congressional Budget Office, *Structuring U.S. Forces after the Cold War: Costs and Effects of Increased Reliance on the Reserves* (September 1992), p. 8; and Department of Defense, "SAR Program Acquisition Cost Summary," June 30, 1997.

a. Numbers may not add to totals because of rounding. Navy active strength would decline from 369,000 to 320,000; Air Force strength from 339,000 to 330,000.

could permit some further reductions in the size of the U.S. warship fleet.

My $15 billion estimate of annual savings reflects a conservative estimate of how these and other cuts in forces and weaponry would affect DoD's budget picture. It assumes that new initiatives in strategic lift and pre-positioning (see below) will be completed quickly, driving short-term costs up. In addition, it takes credit only for savings that are clearly associated with individual military units and fails to

capture savings in Pentagon "overhead" costs. That $15 billion esti-
mated cost reduction reflects only 5.5 percent of DoD's projected
spending level, yet proposed force cuts are closer to 8 percent.[1] Actual
savings might be greater.[2]

Savings are unlikely to be proportional to cuts in the size of the
force, however. Research and development of a new system need to
be completed even if fewer systems are purchased. Unit production
costs are usually greater if fewer systems are purchased. Many intelligence costs, as well as some central costs, are determined more by the
number of foreign targets and the range of potential DoD missions
than by the size of U.S. forces. Some support costs, like airlift, may
depend less on the overall size of the military than on the anticipated
needs for surge deployments (which may not decline as forces do).
Still, it is possible that annual savings from this Desert Storm plus
Desert Shield plus Bosnia strategy could eventually exceed my conservative $15 billion estimate.

RETHINK THE "ONE SIZE FITS ALL" APPROACH TO THEATER WAR

Although not unreasonable, the Pentagon's current insistence on
being able to fight and quickly win two nearly simultaneous wars of
the scale of Desert Storm is more than we really need. The Bottom-Up
Review (BUR) determined that each of these conflicts would involve
a foe comparable in weaponry to 1990 Iraq and would require four to
five Army divisions, four to five Marine brigades, ten Air Force wings,
and four to five Navy aircraft carrier battle groups—together with the
forces of local allies in each theater—to win.

The fundamental logic of the two-war strategy is that the United
States would never want to open a window of opportunity to a potential second aggressor by using most of its forces to confront a first
aggressor. But this apparently prudent logic is excessively cautious, in

1. I am indebted to Jerome Bracken for illuminating conversations on this point.
2. See, for example, the calculations by William W. Kaufmann in John D.
Steinbruner and William W. Kaufmann, "International Security Reconsidered," in
Robert D. Reischauer, *Setting National Priorities: Budget Choices for the Next
Century* (Brookings, 1997), pp. 178–79. Kaufmann does use separate categories
for lift and intelligence and communications, but otherwise assumes DoD costs are
proportionate to force structure.

light of the strong U.S. deterrent postures in key regions, the declining caliber of likely military threats, and various improvements in U.S. capabilities. Moreover, as Senator John McCain points out, the marginal benefits of a robust two-major-war strategy are likely to be outweighed by the damage to military readiness or other consequences of trying to maintain a force structure that is too large in light of likely fiscal constraints.[3]

Still, we know that simultaneous crises can occur in Korea and the Persian Gulf, as events of 1994 in particular underscored. That summer, North Korea threatened to end a commitment to allow its nuclear reactors to be monitored, and Saddam sent troops southward toward Kuwait. For that reason, casual recommendations that the current two-war strategy be downgraded as a basis for U.S. force planning are generally unhelpful unless they wrestle directly with the problem of simultaneous crises or conflicts.[4] The National Defense Panel recently fell into this trap, stating on the one hand that the two-theater construct was an "inhibitor to reaching the capabilities we will need in the 2010–2020 time frame" but equivocating that it "remains a useful mechanism today." That effort to have it both ways, and avoid the tough question about what exactly could or should replace the current two–Desert Storm capability without causing serious potential harm to U.S. interests, understandably did not receive a positive reception from Defense Secretary William S. Cohen.[5]

Just because a two-crisis capability is needed does not mean that the country requires a two–Desert Storm force posture. Rather than plan seriously to wage two regional wars at once, the United States should think in terms of waging a single war while robustly deterring, and in a worst case effectively initiating, a second. Deterrence requires substantial military capability, so even this alternative approach would be

3. John McCain, "Strategy and Force Planning for the 21st Century," *Strategic Review* (Fall 1996), p. 10.

4. For a concurring view, see Paul K. Davis and Richard L. Kugler, "New Principles for Force Sizing," in Zalmay M. Khalilzad and David A. Ochmanek, *Strategy and Defense Planning for the 21st Century* (Santa Monica, Calif.: RAND Corporation, 1997), pp. 95–99.

5. National Defense Panel, *Transforming Defense: National Security in the 21st Century* (Arlington, Va.: December 1997), p. 23; and Douglas Berenson and Roman Schweizer, "Cohen's Draft Response to NDP Reasserts Merits of Two-War Approach," *Inside the Pentagon*, vol. 13 (December 4, 1997), p. 1.

demanding. If deterrence somehow failed in the second theater while full-scale war was already under way in the first, the United States would need viable military options. But those options do not need to cover the full gamut of possible operations in both places at once.

What U.S. military capability would be likely to provide a robust deterrent force? First, the permanent presence of U.S. military personnel in places like South Korea, Kuwait, and Saudi Arabia should be retained. It shows a would-be aggressor that American lives would almost certainly immediately be lost, and thus the American people and government instantly engaged, in any war the aggressor initiated.

The U.S. military must also be able to deploy enough force to set up a robust defense perimeter, stop enemy attacks, and begin a counterattack. Moderate-sized deployments involving rapidly deployed airpower and ground forces on the scale of the 200,000-strong Desert Shield operation could perform this task. If deployed quickly, they could probably prevent initial losses of allied territory; they could certainly stanch any further losses once in place.

The ground units that would make up such a 200,000-strong Desert Shield–like force could protect key infrastructure, dig in to impede enemy offensives, and ensure that land would not be lost even if poor weather or other conditions limited the effectiveness of U.S. airpower in the early going. They could also channel enemy ground forces into certain sectors and slow their rate of movement, increasing their vulnerability to airpower. U.S. airpower would, under most weather conditions, prove devastating to enemy forces. This Desert Shield–like force would generally not be adequate to undertake a full-fledged ground counteroffensive or counterinvasion itself. But it would allow powerful attacks to be made against enemy armor, infrastructure, and economic assets.

Some would argue that deterrence is a concept that one cannot even attempt to apply against the likes of Saddam Hussein or the North Korean (DPRK) regime. They are wrong. Deterrence can admittedly be more difficult to achieve against a risk-taker like Saddam or a desperate regime like that of the DPRK, and in an extreme case it could fail.[6] That is why the United States needs to retain a full-fledged

6. I thank Richard Haass for this point. See also Michael O'Hanlon, "Stopping a North Korean Invasion: Why Defending South Korea Is Easier Than the Pentagon Thinks," *International Security*, vol. 22 (Spring 1998).

warfighting capability for a single major conflict and deterrence, including a strong initial response capability for a second. But clear threats and lines in the sand are usually not lost on individuals who have proven their street smarts by gaining power and keeping it.[7]

Although both North Korea and Iraq have tested U.S. will in the recent past, they have wisely displayed a strong penchant for avoiding large-scale combat when American resolve to fight back against them was clear. It was not clear in early 1950 or early to mid-1990, respectively, but it is clear now. We should not lose sight of the distinction between testing resolve and provoking war. Testing resolve is a dangerous game to play. It is unlikely to lead to actual combat, however, unless at least one involved party really wants combat. As Geoffrey Blainey has argued compellingly, few wars happen by accident.[8]

The United States should plan to rapidly deploy Desert Shield–like forces in both southwest and northeast Asia if needed. It should also be able to expand one of those deployments to a much larger capability should that prove necessary, either because somehow an initial halting operation does not succeed or because allied forces choose to counterinvade the aggressor's territory and perhaps overthrow its government. To do these things, the United States should retain a Desert Storm plus Desert Shield capability. In addition, it should keep a modest amount of additional forces for operations other than war, yielding an overall force posture with a Desert Storm plus Desert Shield plus Bosnia capability.

In 1993 such a capability was considered during Pentagon discussions leading up to the BUR force. Nicknamed "win/ hold/win" by its proponents, it was derided as "win/hold/oops" by its detractors and ultimately dismissed as too risky. But the detractors, and even the proponents, did not give its capabilities enough credence. Rather than be coined a "win/hold/win" force, it should be viewed as a "counterinvade/hold-and-punish/counterinvade" capability. Indeed, under many circumstances the "hold-and-punish" capability would be adequate to achieve victory and bring hostilities to a close. Under many circumstances it would not be enough to achieve an unconditional surrender.

7. For a concurring view, see Janne Nolan and Mark Strauss, "The Rogues' Gallery," *Brown Journal of World Affairs*, vol. 4 (Winter–Spring 1997), pp. 32–35.

8. Geoffrey Blainey, *The Causes of War* (Free Press, 1973), pp. 127–45.

But in some cases it might be. Notably, given the strength of the South Korean (ROK) military and the likely attrition that would result to North Korean units during any initial DPRK attack, a U.S. Desert Shield–like force might include enough American ground units that when teamed with the South Koreans they could conduct a full-scale counterinvasion.[9]

This military strategy, with its focus on rapid responsiveness, embodies a slightly greater relative emphasis on airpower than does current U.S. defense strategy. Because of the requirements imposed by peacekeeping, overseas presence, and a possible single Desert Storm–like operation, however, the increased focus on airpower is of limited scope. Moreover, as explained further in chapter 4, it does not provide blanket support for all Air Force modernization programs.[10] But overall strategic and technological trends do allow the United States to place a somewhat greater emphasis on Air Force weaponry than in past decades.

The Potency of Modern Airpower and Precision Munitions

The main reason that a smaller force, if deployed quickly, can be so effective is that exposed armor is highly vulnerable to detection and attack by today's U.S. military. For example, data from the Gulf War experience indicate that a land-based contingent of approximately 500 U.S. attack aircraft could probably destroy 1,000 moving enemy vehicles per 24-hour period—each flying an average of two sorties a day and delivering on average four munitions each with at least 25 percent probability of destroying an armored target. Those destruction rates can be achieved by existing munitions: laser-guided bombs; Maverick air-to-ground missiles guided by laser, TV, or infrared sensors; and helicopter-delivered Hellfire, currently guided by infrared or laser sensors but soon also to have all-weather radar homing capa-

9. See O'Hanlon, "Stopping a North Korean Invasion."

10. For strategic concepts that place considerably more emphasis than I do on airpower, see Zalmay Khalilzad and David Ochmanek, "Rethinking US Defence Planning," and Benjamin S. Lambeth, "The Technology Revolution in Air Warfare," *Survival*, vol. 39 (Spring 1997), pp. 43–64 and 65–83, respectively; and Barry M. Blechman and Paul N. Nagy, *U.S. Military Strategy in the 21st Century* (Arlington, Va.: IRIS Independent Research, 1997).

bility. Those attrition rates would virtually grind a ten-division thrust to a halt in just a few days.[11]

Even higher attrition rates might result when moving armor is attacked, as in the battle of Al Khafji. Moving armor also is easier to detect with radar-imaging aircraft, increasing the odds that attack planes would consistently have targets to shoot at with their high-lethality ordnance.[12]

Anticipated improvements in the next generation of munitions will probably increase the vulnerability of moving armor further. The new weapons consist largely of autonomous submunitions of the "fire and forget" variety that do not even require the aircraft launching the munitions to precisely target the enemy armor under attack. In some cases they can be released by airplanes from standoff range and then be "dispensed" to attack targets once a missile aided navigationally by global positioning system (GPS) satellites takes them to the general vicinity of enemy forces.[13] In regard to this specific military mission—using mod-

---

11. The CIA found that, during just the air war, some 1,135 Iraqi tanks and 827 armored personnel carriers were destroyed by coalition aircraft. A comparable number of artillery were probably destroyed as well, making for a rough total of perhaps 3,000 major armored weapons destroyed by aircraft before the ground war. Although no definitive evaluation of the relative effects of precision munitions versus unguided munitions is possible, pilots generally reported much greater effectiveness with guided weapons. Perhaps another 1,000 armored vehicles were destroyed from the air during the ground war. So air-delivered precision munitions, of which almost 15,000 were dropped during the war—apparently not many more than 5,000 against enemy armor—probably destroyed 2,000 to 3,000 weapons over the course of a comparable number of aircraft sorties. See General Accounting Office, *Operation Desert Storm: Evaluation of the Air Campaign*, NSIAD-97-134 (June 1997), pp. 25, 147, 181, 211, 218–19; Timothy M. Laur and Steven L. Llanso, *Encyclopedia of Modern U.S. Military Weapons* (New York: Berkley Books, 1995), pp. 267–74; Department of Defense, *Conduct of the Persian Gulf War: Final Report to Congress* (April 1992), pp. T18–T19, T182–T186; Thomas A. Keaney and Eliot A. Cohen, *Gulf War Air Power Survey Summary Report* (Government Printing Office, 1993), pp. 103–15, 197–203; and Michael O'Hanlon, *Defense Planning for the Late 1990s: Beyond the Desert Storm Framework* (Brookings, 1995), pp. 55–72.

12. Keaney and Cohen, *Gulf War Air Power Survey Summary Report*, p. 109; and Thomas G. Mahnken and Barry D. Watts, "What the Gulf War Can (and Cannot) Tell Us about the Future of Warfare," *International Security*, vol. 22 (Fall 1997), pp. 155–56.

13. For a good summary of efforts underway, applicable to tactical as well as bomber aircraft, see David Mosher, *Options for Enhancing the Bomber Force* (Congressional Budget Office, July 1995), pp. 21–36.

ern munitions against exposed enemy armor—the often overrated "revolution in military affairs" may indeed have something to offer. (For a broader discussion of the "RMA" concept, see chapter 4.)

The Importance of Preparation and Speed

To be able to deploy Desert Shield–like forces quickly enough to preclude territorial losses in both Korea and the Persian Gulf, U.S. military planners should put even greater emphasis on rapid response. The Clinton administration deserves considerable credit for its initiatives in improving the responsiveness of U.S. forces to future crises or conflicts. One of the troubling dimensions of Desert Shield and Desert Storm, frequently noted by commentators, was the U.S. need to have several months of buildup time even in such a well-prepared theater. That was largely due to a lack of adequate pre-positioning in the theater and adequate surge sealift.

Significant progress has been made in these regards. The payoffs will grow rapidly over the rest of the decade as construction of large, medium-speed roll-on/roll-off ships is completed and more equipment is based in northeast and southwest Asia and afloat overseas (see figures 3-1 through 3-3 and table 3-2). The intent is to be able to deploy a light division, send over troops to marry up with one to two brigades of pre-positioned heavy equipment, and then sealift two full heavy divisions within a month.[14] More Marine forces could begin to arrive in that time as well. A full five-division corps would be delivered in principle within seventy-five days as fast sealift ships completed a second voyage to the region and as slower sealift ships were made serviceable, loaded, and deployed.[15]

Why make the remaining two to three active heavy divisions wait so long for deployment? That plan would present a troubling Achilles'

14. The pre-positioning programs have been expanded considerably and despite some glitches are working well. An Army brigade set of materiel has been added in the Pacific region; a brigade set has also been added in the Middle East, with another due to be in place by 2000; and the reliability of the equipment in these sets is typically at least 90 percent. See General Accounting Office, *Overseas Presence*, NSIAD-97-133 (June 1997), p. 22; and General Accounting Office, *Afloat Prepositioning*, NSIAD-97-169 (July 1997), pp. 3– 5.

15. Rachel Schmidt, *Moving U.S. Forces: Options for Strategic Mobility* (Congressional Budget Office, February 1997), pp. 29, 79.

FIGURE 3-1. *Sealift Capacity under the Administration's Plan, 1996–2001*[a]

Millions of square feet of cargo space[b]

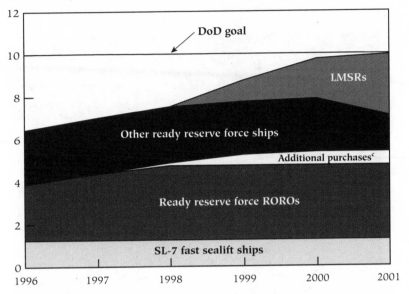

DoD goal

LMSRs

Other ready reserve force ships

Additional purchases[c]

Ready reserve force ROROs

SL-7 fast sealift ships

Source: Rachel Schmidt, *Moving U.S. Forces: Options for Strategic Mobility* (Congressional Budget Office, 1997), p. 24, based on data from the Department of Defense.

a. Each LMSR (large, medium-speed roll-on/roll-off ship) can carry about 300,000 square feet of cargo and is equipped with its own set of cranes; each RORO (roll-on/roll-off ship) holds about 100,000 square feet of cargo; and each SL-7 holds about 150,000 square feet of cargo plus 188 20-foot-equivalent containers (the eight SL-7s owned by DoD are enough to carry a heavy division).

b. Millions of square feet represent the total sealift capacity measured by the sum of all stowage space in a single sailing of all available sealift ships.

c. The DoD goal assumes additional purchases of five more ROROs or the equivalent.

heel if two major crises developed within days or weeks of each other. A better approach would buy enough lift to deploy two Desert Shield–like forces to two separate places within a month, rather than just one as current plans envision.

The Pentagon could build more large, high-value, LMSR (large, medium speed roll-on/roll-off) class vessels for this purpose. But these ships are expensive, put many U.S. eggs in one basket, and could clog up many of the world's ports. Instead, DoD should purchase somewhat slower yet much cheaper roll-on/roll-off ships like those now in the ready reserve. Whereas twenty more LMSRs could suffice for two

FIGURE 3-2. *Strategic Airlift Fleet under the Administration's Plan, 1996–2007*[a]

Millions of ton-miles per day[b]

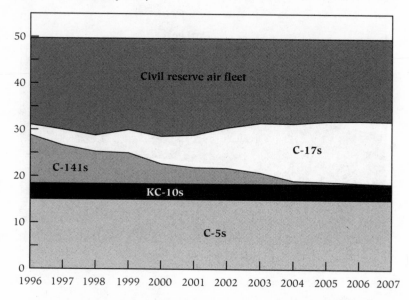

Source: Schmidt, *Moving U.S. Forces*, pp. 11–13, based on Department of the Air Force, *1997 Air Mobility Master Plan* (Scott Air Force Base, Ill.: Air Mobility Command, October 11, 1996).

a. Both the C-5 and the C-17 can carry outsize cargo, such as M-1 tanks. The C-5 has an average load of 65 tons, and the C-17 of 45 tons. The C-141, with a payload of 23 tons, is the primary plane used to air-drop personnel and supplies. The KC-10 and the KC-135, used as tankers and airlift, are military versions of civilian aircraft that need special equipment for loading and unloading. The KC-10 has an average payload of 40 tons; the KC-135 has an average payload of about 10 tons. The CRAF fleet carries bulk supplies and troops only.

b. The standard unit of measure of theoretical airlift capacity, which takes into account both the weight of cargo and the distance over which it must be carried. For instance, 49 MTM/D represents the ability to move 49,000 tons of cargo over 1,000 nautical miles in a day's time (including the return trip to home base).

more divisions' worth of combat forces and some initial supplies, this approach might require fifty smaller ships. But at $35 million per smaller ship, that would amount to less than $2 billion, in contrast to the likely cost for twenty LMSRs of about $6 billion.[16] This type of initiative could have the added benefit of softening the pain that U.S. shipyards would suffer from the cutbacks in DDG-51 destroyer production advocated elsewhere in this study (see chapter 4).

16. Ibid., pp. 26–29.

FIGURE 3-3. *Amphibious Lift Capacities for the U.S. Marine Corps, 1993–2015*[a]

Marine expeditionary brigade equivalents

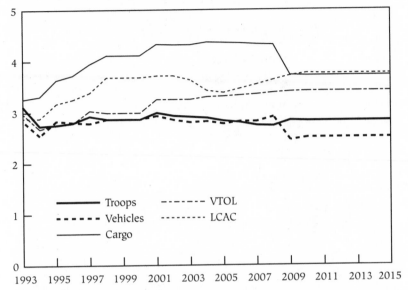

Source: U.S. Marine Corps, based on administration and Marine Corps plans.

a. VTOL (vertical take-off and landing) aircraft include Harrier jets and helicopters; LCAC = landing craft, air cushion. A marine expeditionary brigade, roughly 15,000 troops in all, includes some 17 tanks, 30 155-millimeter artillery, 85 other armored vehicles, 90 helicopters, 40 Harrier jets, and 24 F/A-18 jets. See Michael Berger, *Moving the Marine Corps by Sea in the 1990s* (Congressional Budget Office, 1989), p. 15.

Improvements are also needed in the military's ability to move supplies around within the United States. During Desert Shield and Desert Storm, Army supplies typically took over a month to reach their U.S. port or airfield of debarkation. Automation and computerization are the keys to improving this antiquated logistics system and one of the areas where commercial practices can be most directly adopted by DoD. It is beyond the scope of this study to estimate the costs of expediting these efforts, but they appear modest and likely to be balanced out by savings from reduced spare parts inventories within a short time.[17]

17. See "RAND: Army's 'Slow and Unreliable' Distribution Process Can Be Improved," *Inside the Pentagon*, January 8, 1998, pp. 7–8.

TABLE 3-2. *Current and Future Pre-positioning Sites under the Administration's Plan*[a]

Sites	Current capabilities	Future added capabilities
Europe		
Benelux countries	2 Army POMCUS brigade sets	. . .
Norway	1 Marine MEB set	. . .
	1 Army artillery battalion set	. . .
Italy	1 Army brigade set	. . .
Mediterranean	Marine MPS squadron 1	Expand Marine MPS
	1 Air Force ammunition ship	squadron by 1 ship
Asia		
South Korea	1 Army brigade set	. . .
Guam and Saipan	Marine MPS squadron 3	. . .
	Army combat-support and combat-service-support assets	
Persian Gulf and Indian Ocean		
Persian Gulf	1 Army brigade set in Kuwait	Complete 1 Army brigade set plus division base set in
	1 Army battalion task force set in Qatar	Qatar
	4 Air Force bare base sets	Possibly another heavy-brigade set elsewhere
Diego Garcia	1 Army brigade set plus theater-opening support	Expand Army APA from 870,000 square feet to 2
	3 (joint-service) DLA fuel ships	million square feet of cargo
	2 Air Force ammunition ships	Expand Marine MPS squadron by 1 ship
	Marine MPS squadron 2	. . .

Source: Rachel Schmidt, *Moving U.S. Forces: Options for Strategic Mobility* (Congressional Budget Office, February 1997), pp. 35–42.

a. Table does not include equipment that is intrinsic to deployed forces. Benelux = Belgium, the Netherlands, and Luxemburg; POMCUS = pre-positioning of materiel configured to unit sets; MEB = marine expeditionary brigade; MPS = maritime pre-positioning ships; DLA = Defense Logistics Agency; APA = Army pre-positioned afloat program.

It would also behoove U.S. policymakers to buttress pre-positioning in southwest Asia even more. As shown in table 3-2, the United States now has slightly more than a division equivalent of heavy equipment in the Persian Gulf/Diego Garcia region, with plans to increase it to roughly 1.5 division equivalents. But even more insurance is desirable, in keeping with the Desert Shield model. Being able to get two full heavy divisions to the region within a few days, and also to send a division of light forces from the United States in that time as well, would be desirable. So would storing more Air Force precision munitions in the area. The simplest and least expensive way to address the ground-component side of things would be to move the two brigades of POMCUS (pre-positioning of material configured to unit sets) equipment still in Europe to the region. Given the Saudi regime's reluctance to increase the visibility of the U.S. military in its country, that POMCUS might have to be afloat or based in the southeastern part of the Arabian peninsula. That could render it more vulnerable to having its arrival delayed in wartime by enemy submarine or mine threats, as discussed below. But such basing arrangements could have advantages too, seriously complicating any future adversary's efforts to destroy POMCUS stocks in the early phase of a future attack. Since the combat equipment is already available, the annual cost over ten years could be held to $200 million.

Airlift should also be increased, with the goal of deploying enough troops, aircraft, and other supplies for two Desert Shield–like operations within a month of the beginning of a crisis. With roughly fifty 747-class aircraft and fifty KC-10 aircraft added to its planned inventory, the Air Force could achieve that goal. It could achieve a sixty million-ton-miles per day (MTM/D) deployment rate rather than the current fifty MTM/D capability. The associated total cost would be about $20 billion.[18] Once paid over the next decade or so, and once C-17 procurement funding is completed in the year 2003 with the 120th aircraft, more modest airlift costs of roughly $3 billion might go toward refurbishment of the C-5A fleet to keep it in service over an extended period.[19]

18. O'Hanlon, *Defense Planning for the Late 1990s*, pp. 55–65.
19. Jeffrey Record, "After the C-17: Coming Airlift Realities and Choices," *Armed Forces Journal International* (December 1996), pp. 32–33; and Schmidt, *Moving U.S. Forces*, p. 17.

Some see the B-2 as the way to handle rapid armored thrusts in distant regions, without requiring dependence on overseas bases that may prove either unavailable or vulnerable. In theory those concerns argue in favor of the B-2. But bombers like the B-2, though highly touted for their alleged capabilities to stop rapid armored advances, have no such capability now, and it is not clear when and if they will ever acquire it. If they do acquire that capability, it will be largely thanks to the quality of future munitions. And if those munitions prove as good as advertised, they may not require delivery by a penetrating bomber. A combination of stealthy, lower-value reconnaissance platforms with standoff munitions might be a much safer and more economical way to accomplish the same goal (see chapter 4).

Korea

Another consideration adds credence to the Desert Storm plus Desert Shield plus Bosnia philosophy: the quality of South Korean armed forces and the robustness of the U.S.-ROK defense posture on the Korean peninsula today. It is quite unlikely that North Korea could make significant inroads into South Korea in the early phases of a war, even in the event that it used chemical weapons and attacked during bad weather, when allied aircraft would be less effective against ground targets. Although Seoul would be severely damaged by artillery, missiles, and special forces, it almost certainly could not be taken by DPRK units in their current state. Any counterinvasion of North Korean territory could still be challenging for allied forces. But the time urgency of conducting it would be less great than an operation to liberate South Korea—and the amount of damage that North Korean forces would have suffered in any all-out attack on the ROK might weaken their forces so much that South Korea could handle most of the counterinvasion on its own anyway.[20]

To see why the North Korean battle plan is so unpromising, first consider historical precedent. When armies tried to drive directly through prepared defensive positions in World War II, for example—what North Korea would have to do in a future war on the peninsula—

20. This argument is developed in greater length in O'Hanlon, "Stopping a North Korean Invasion."

they rarely advanced more than four or five kilometers a day. Advance rates were as low as about one kilometer a day in campaigns against well-prepared terrain, such as the allies' attack against Germany around the Siegfried Line when they enjoyed air superiority and significant numerical advantages.[21] Without advancing very rapidly, North Korean forces would become increasingly vulnerable to U.S. tactical aircraft reinforcements and lose whatever protection attacking in bad weather may have given them against U.S. munitions dependent on laser-homing or infrared guidance. They would also probably run out of supplies after a few days, since allied artillery and air attacks as well as mines and road obstacles would prevent significant movement of resupply vehicles down key invasion corridors.[22]

ROK defenses are dense. Most of the South Korean army is deployed across just 250 kilometers of front. The resulting force-to-space ratio of about one division per 10 kilometers is excellent. Modern ground forces are generally designed to cover at least twice that much frontage.[23]

Defenders would be immediately vulnerable to artillery attack at the rate of up to tens of thousands of rounds per minute. But so would attackers—and they would be exposed. The exposed area, and thus the vulnerability, of a soldier on foot or in a jeep or truck would be at

21. Robert L. Helmbold, "A Compilation of Data on Rates of Advance in Land Combat Operations," Research Paper CAA-RP-90-04, U.S. Army Concepts Analysis Agency, Bethesda, Maryland, February 1990, pp. A-278—A-294, A-318—A-322; and Barry R. Posen, "Measuring the European Conventional Balance," in Steven E. Miller, ed., Conventional Forces and American Defense Policy (Princeton University Press, 1986), p. 114.

22. For background, see Frances Lussier, U.S. Ground Forces and the Conventional Balance in Europe (Congressional Budget Office, June 1988), p. 86. The North Koreans would need to send at least 1,000 truckloads of supplies southward daily to support a large armored force. See Joshua M. Epstein, Strategy and Force Planning: The Case of the Persian Gulf (Brookings, 1987), pp. 112–13.

23. William P. Mako, U.S. Ground Forces and the Defense of Central Europe (Brookings, 1983), pp. 36–37; John Patrick Elwood, "Conventional Wisdom: Force-to-Space Considerations and Conventional Arms Control in Europe," senior thesis, Princeton University, 1989, p. 28, based on data in Robert McQuie, Historical Characteristics of Combat for Wargames (Benchmarks), CAA-RP-87-2 (Bethesda, Md.: U.S. Army, Concepts Analysis Agency, July 1988); and Joshua M. Epstein, Conventional Force Reductions: A Dynamic Assessment (Brookings, 1990), p. 58.

least ten times as great as of a soldier in a foxhole.[24] Similar considerations apply to gunfire during close-in battle.

Allied military equipment is also much more capable than that of the DPRK. South Korean Type 88 or K-1 tanks, for example, have detection and targeting sensors similar to those of the U.S. M1 Abrams. They would be firing at even more primitive mixes of Soviet-style tanks than coalition forces confronted in southwest Asia in 1991. North Korea owns T-62s, which entered production in the early 1960s, and earlier vintages. T-62 tanks and similar systems had a mediocre track record twenty-five years ago in the Arab-Israeli wars and would do even worse against modern antitank weapons and tanks.[25]

Allied reconnaissance is excellent.[26] Not only could it detect any large-scale massing of armored vehicles in any weather through platforms like overhead reconnaissance satellites and Joint STARS radar-imaging aircraft, it could also monitor the movement of approaching human beings through devices such as ground radars and infrared detectors.[27]

Some acknowledge that the traditional balance of forces may work in the allies' favor but worry a great deal about North Korea's special forces. But whatever their capacity to wreak terror in places like Seoul, the North Korean commandos' military effectiveness against South Korean defensive lines would probably be quite limited. Airborne assault over the battlefield would generally be suicidal,

24. Comparisons with the Iran-Iraq war are telling: in that conflict, most artillery-caused casualties were exposed soldiers attempting to effect an advance, and most successful infiltrations of minefields were the result of careful probing attacks rather than massed assaults. Both of these results involving armies of roughly comparable technology and training to North Korea's bode poorly for the latter's ability to carry out a successful massive surprise attack. See Anthony H. Cordesman and Abraham R. Wagner, *The Lessons of Modern War,* vol. 2: *The Iran-Iraq War* (Boulder, Colo.: Westview Press, 1990), pp. 433, 445, 447.

25. Stephen Biddle, "Victory Misunderstood: What the Gulf War Tells Us about the Future of Conflict," *International Security,* vol. 21 (Fall 1996), pp. 167–72; and David C. Isby, *Weapons and Tactics of the Soviet Army* (London: Jane's Publishing Company, 1981), pp. 130–50.

26. See, for example, James L. Stokesbury, *A Short History of the Korean War* (Morrow, 1988), pp. 102, 120.

27. See, for example, *Jane's Weapon Systems 1988–89* (Alexandria, Va.: Jane's Information Group, 1988), pp. 282–84.

given the allies' control of the air. Tunnel assault would be slow, vulnerable to counterattacks, and unable to penetrate deeply into ROK defenses, given the limited length of the tunnels.[28] Infiltrations over land and by submarine would generally limit troop numbers to dozens or hundreds—a potent capability against only relatively undefended targets.

What about weapons of mass destruction? The primary concern is chemical weapons. Any nuclear weapons the DPRK may have are unlikely to be of great benefit on the battlefield.[29] A surprise nuclear strike that luckily got through defenses might be able to create a hole of perhaps one to two kilometers radius in South Korean defensive lines. But that would open up DPRK forces to the prospect of rapid nuclear retaliation, quite likely in the same sector where they had attacked South Korean defenses—probably denying North Korea the opportunity to exploit the gap that had been created. North Korean biological weapons would probably be too difficult to employ in a militarily effective manner that did not also threaten DPRK troops and would in any case be of limited benefit in the early days of a surprise attack.[30] Any use of biological weapons would also raise the prospect of allied nuclear retaliation. Chemical weapons might too, but since they have been used before in war and can be kept largely isolated from civilian populations, North Korea might hope that it could employ them yet escape nuclear retaliation (for example, against its troops massed in invasion corridors near Seoul).

The main immediate concern would be what North Korea could do to front-line allied forces in the early stages of an attack. In the critical early stages of any North Korean attack, those front-line ground forces would be the key to a successful defense. (Allied operations at airbases and ports could be slowed but not prevented by chemical attacks, and in any case Japanese air bases would probably be unscathed and available as well. But given my assumption that the

28. Nick Beldecos and Eric Heginbotham, "The Conventional Military Balance in Korea," *Breakthroughs* (Spring 1995), p. 5.

29. It is more dubious than most realize that North Korea ever extracted enough plutonium to make a bomb. See Leon V. Sigal, *Disarming Strangers: Nuclear Diplomacy with North Korea* (Princeton University Press, 1998), pp. 92–95.

30. Defense Intelligence Agency, *North Korea: The Foundations for Military Strength* (October 1991), pp. 60–62.

DPRK would attack during bad weather, airpower's effectiveness would be limited even without chemical attacks.)[31]

A North Korean attack against front-line units would probably use nonpersistent or "high volatility" chemical agents. Use of persistent chemicals would likely cause very serious troubles to allied forces, who probably do not have adequate protection and decontamination equipment against them.[32] But it would also prevent North Koreans, who likely lack sufficient numbers of good protective suits and would have trouble covering several kilometers of land on foot while suited up even if they owned them, from exploiting any holes in a timely fashion.[33]

DPRK chemical attacks using nonpersistent agents, though dangerous, would generally require only that allied units use gas masks (not suits)—and then only for a relatively short time. Although wearing gas masks is always difficult, it is far less taxing when one is manning a fixed defensive position than when on the assault. On the other hand, North Korean military leaders could face a difficult choice if attempting to profit from their gas attack promptly: either force their own troops to breathe heavily through gas masks while attacking or leave them vulnerable to gas that might persist longer than expected and be carried in unexpected directions by the wind. South Korea could lose some troops in the early stages of a chemical attack, if they were surprised. But most troops keep their masks nearby at all times, and warning detectors are abundant. On the whole, the chemical threat against front-line, dug-in troops is significant but is unlikely to be the determining factor in the outcome of any battle.[34]

31. For further discussion, see O'Hanlon, "Stopping a North Korean Invasion."

32. For a limited discussion of the needs to improve U.S. chemical protection and decontamination capabilities, see William S. Cohen, *Report of the Quadrennial Defense Review* (Department of Defense, May 1997), pp. 24, 49. For what little information is publicly available on other countries' detection, protection, and decontamination capabilities, see Terry J. Gander, ed., *Jane's NBC Protection Equipment* (Alexandria, Va.: Jane's Information Group, 1995).

33. Militaries short on training can have serious problems in this, as in many other, facets of warfare. For example, early in the Iran-Iraq war, Iraq killed more of its own soldiers than of its enemy when employing mustard gas. See David Kay, Ronald F. Lehman, and R. James Woolsey, "First the Treaty, Then the Hard Work," *Washington Post*, April 13, 1997, p. C7. See also Victor Utgoff, *The Challenge of Chemical Weapons: An American Perspective* (St. Martin's Press, 1991), pp. 148–88.

34. Utgoff, *Challenge of Chemical Weapons*, pp. 162–70.

This assessment is reinforced by reference to other wars, in which chemical weapons were generally not a dominant cause of casualties. For example, in the Iran-Iraq war, they were responsible for less than 5 percent of all casualties.[35] Since allied forces have several times the firepower needed to halt a North Korean attack, a degradation of several percent in their strength should not change the basic course of battle.

Although the official Pentagon image of war in Korea appears to be pessimistic, at least one important think tank associated with the Department of Defense shares my more positive assessment of the likely outcome of the initial defensive effort north of Seoul. Specifically, the Army's Concepts Analysis Agency, in a draft working paper, characterizes the initial fight as "primarily a South Korean ground force fight" with U.S. airpower also playing a role deemed "critical to success"—and the overall effort is expected to stop the North Korean offensive north of Seoul. U.S. reinforcements would be deployed principally to help mount a counterinvasion that would go north of Pyongyang.[36]

Better Safe than Sorry

Even with additional improvements in strategic lift and pre-positioning, and even in light of the trends favoring South Korea over North Korea, a forward-defense strategy could still prove insufficient. At the level of political and strategic uncertainty, a key allied regime could be overthrown from within, war could occur in an unexpected place, or conflict could break out in a murky way that made it difficult for the United States and regional allies to decide quickly on an appropriate course of action. At the more technical military level, bases or pre-positioned military stocks could be destroyed in a surprise commando attack or through the use of weapons of mass destruction. Ships carrying pre-positioned equipment could be impeded from arriving in their ports of destination by the threat of mines or enemy submarines, either of which could take weeks to clear out under a

35. Cordesman and Wagner, *Lessons of Modern War*, vol. 2, p. 518.

36. U.S. Army Center for Strategy and Force Evaluation, "Strategy-Based Campaign Analysis: An Assessment of Alternative Force Structures," Draft Working Paper, 1997.

worst-case assumption.[37] U.S. and allied forces might also decide to respond to an attack by invading an adversarial country to overthrow its regime or confiscate weapons of mass destruction, necessitating use of occupation forces. So a warfighting strategy focused nearly exclusively on the halt phase would be inadequate.

As suggested above, the concept of a Desert Storm plus Desert Shield plus Bosnia framework seems most sound. In other words, DoD could first attempt to rely on rapid responsiveness and potent airpower, together with allied capabilities and smaller U.S. ground forces, in the event of major regional war. Its Desert Shield–like force package would have roughly the full airpower complement that the BUR and QDR envision for major theater wars, but somewhat less than half as many ground forces and a total troop size of only about 200,000. The United States would, however, retain the prompt ability to augment one of those smaller Desert Shield–like deployments into a much larger Desert Storm–like capability—or a major theater war force, in the parlance of the QDR.[38]

Under this approach, the requirement for a single Desert Storm operation should be calculated conservatively in case a war proves considerably more demanding than anticipated. As noted in chapter 1, the QDR's "building blocks" for regional warfighting already do that to an extent. Various DoD models and simulations indicate that a force only about 80 to 90 percent as large as today's might well be able to handle two overlapping regional conflicts of the Desert Storm variety.

A further warfighting insurance margin exists in the form of the large Army National Guard combat force structure, which even if cut back significantly would retain many more units than are now built into war plans (as explained in more detail in chapter 5).[39] Notably, at a minimum its fifteen enhanced separate brigades should be retained

37. See "Cohen Directs Navy to 'Reevaluate' Spending Cuts to Mine Warfare Programs," *Inside the Navy*, December 15, 1997, p. 1; and Owen Coté and Harvey Sapolsky, *Antisubmarine Warfare after the Cold War* (Cambridge, Mass.: MIT Security Studies Program, 1997), p. 13.

38. For a generally similar view about warfighting requirements, see McCain, "Strategy and Force Planning for the 21st Century," p. 10.

39. See Richard L. Kugler, "Nonstandard Contingencies for Defense Planning," in Paul K. Davis, ed., *New Challenges for Defense Planning* (Santa Monica, Calif.: RAND Corporation, 1994), pp. 165–96.

in the future. The enhanced brigades would also prove a useful capability for the unlikely situation in which two or more difficult peacekeeping missions placed simultaneous demands upon U.S. forces. One or more could be called up either to assist in a peacekeeping mission or to begin intensive training as a fill-in for active-duty forces so employed.

Taking these various considerations into account, the Desert Storm plus Desert Shield plus Bosnia force-sizing approach would permit reducing active-duty military strength by nearly 10 percent by comparison with the QDR, largely in ground and naval forces.

Make Units "Thinner," Not Fewer in Number

To effect this troop reduction, it makes more sense to reduce the size of most types of units rather than their number. Most notably, preserving an active Army with ten divisions, an active Marine Corps with three, and an Air Force with a total of twenty wings would help stabilize the U.S. military after a decade of downsizing and avoid creating the perception here and abroad that forces were continually being cut and eroded. The modest reductions in the size of units could be justified as not only possible but desirable based on recent and expected trends in military technology. Improvements in the force that have occurred over the 1990s—such as a major increase in the number of Air Force attack aircraft with LANTIRN infrared targeting, introduction of Joint STARS and UAVs into the force, and further purchases of systems like multiple launch rocket systems as well as the advent of "smart" homing submunitions—already could provide a good "microlevel" justification for such changes.

Keeping today's overall force architecture would also allow a sufficient rotation base for the frequent operations other than war that have strained the U.S. military since the cold war ended. Even if their utility in many major theater war scenarios can be questioned, units like the Army's 10th Mountain Division have been exceptionally busy and useful in places like Somalia and Haiti. Nor are they irrelevant to major theater wars, especially if urban combat proves necessary, military occupations are carried out, or terrain is rugged.

The idea of reducing the size of standard units has been taken seriously by a number of top military leaders already. Former Air Force chief of staff Ronald Fogleman once suggested that future Air Force fighter wings might contain only sixty rather than seventy-two primary aircraft. That thinking reflects both increased performance parameters like speed and stealth, as well as expectations about high reliability and sustainable sortie rates for future planes.

A number of Marine Corps officials envision a more mobile and agile future force with battalions perhaps only half the size of today's, more logistics and firepower based at sea, and therefore less need for large security elements ashore to protect rear areas.[40] (Indeed, the Cheney/Powell base force envisioned maintaining three Marine divisions with only 159,000 active-duty troops rather than the 172,000 planned under the QDR.) Marines are not yet sure that their future ground forces can be smaller. But the logic of their future view of warfare—conducted more from the sea, focused more on urban operations, less apt to involve large beachheads or bases ashore—means at a minimum that the Marines can give up their fixed-wing airpower. Marines should provide their own helicopter and other short-and-vertical takeoff capability but scale back systems that they would not be operating from ships or small bases on land. Fixed-wing airpower could be provided by Navy or Air Force units when needed and feasible.[41]

The Navy, to its credit, has already reduced substantially the number of escort vessels accompanying a standard carrier battle group in light of the reduced threat to carriers (and greater capability of Aegis-class escorts). It is now also displaying open-mindedness to a concept of crew rotation, discussed below, that could greatly increase the efficiency with which the routine peacetime presence mission is conducted.

In the Army, some top officials, such as former assistant vice chief of staff Lieutenant General Jay Garner, have spoken favorably of reducing the size of key units as well.[42] However, what is most notable

40. U.S. Marine Corps, *Concepts and Issues 97* (1997), pp. 38–39; and "Services Study How Technology May Permit Force Reductions," *Inside the Pentagon*, March 21, 1996, p. 3.

41. See Statement of Martin R. Steele and Thomas L. Wilkerson before the Military Personnel Subcommittee of the House National Security Committee, January 29, 1998; and Merrill A. McPeak, *Presentation to the Commission on Roles and Missions of the U.S. Armed Forces* (Department of Defense, September 1994), pp. 69, 91–93.

42. Greg Caires, "Garner: Army Could Reduce the Size of Its Divisions," *Defense Daily*, February 25, 1997, p. 285.

FIGURE 3-4. *Changes in Weight of Various Army Combat Units, 1987–94*

Thousands of tons[a]

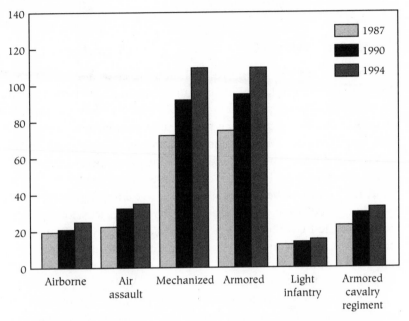

Source: Schmidt, *Moving U.S. Forces*, p. 82. Based on data from the Military Traffic Management Command.
a. The weight of each unit includes accompanying supplies, equipment, and ammunition.

is the trend toward increasingly heavy divisions—every major type of unit is now at least 10 percent heavier than at the time of the Gulf War—even as talk about the need for a more deployable future force becomes popular within the service (see figure 3-4).[43]

Rather than view the goals of increased mobility and decreased weight as distant aspirations for the "force after next," it is time to start pursuing them with the current force. After all, the year 2010—the year of reference for former chairman of the joint chiefs of staff John Shalikashvili's vision of future warfare—is not distant, especially since DoD acquisition programs and force structure changes generally take at least five to ten

43. See U.S. Army Program Analysis and Evaluation Directorate, *America's Army: Projecting Decisive Power* (1995), p. 13; and Schmidt, *Moving U.S. Forces*, p. 82.

years to carry out.[44] The imperative to get going is greatest for the Army, which, as noted, has been going in just the wrong direction, and which could afford to be somewhat smaller under the Desert Storm plus Desert Shield plus Bosnia construct proposed here.[45]

An Insurance Policy: NATO and Out-of-Area Operations

Although not needed to allow the United States to adopt a Desert Storm plus Desert Shield plus Bosnia force posture, expanded allied capability for operations in places like the Persian Gulf would provide an added element of insurance to any such approach. More important, it would allow Western allies that share most U.S. global interests to share also in the risks and costs—including the blood costs—of any future large-scale military operation outside Europe. This issue is clearly not within the immediate purview of Pentagon planners, but its political and strategic importance merits some discussion here.

Neither the BUR nor the QDR assumed any specific military contribution to wars in the Persian Gulf or Korean peninsula from the major NATO allies. Since our allies' defense policy and budget decisions are not made in Washington, the tendency among American defense planners is often to accept that assumption and move on to a tunnel-visioned focus on U.S. forces. But this assumption deserves scrutiny, in light of the common Western interests at stake in places like the Persian Gulf.

Some progress in NATO's mechanics has been achieved in recent years. For example, the so-called Combined Joint Task Force (CJTF) concept has been established. It allows a coalition of willing alliance members to use NATO assets in the conduct of operations that, while not mandated by the obligatory mutual self-defense clause (Article V) of NATO's charter, nevertheless appear important to a number of countries. Also, the four-nation Eurocorps has institutionalized joint training and deployment planning for forces provided by France,

---

44. See John Shalikashvili, *Joint Vision 2010* (Department of Defense, 1995), pp. 13–20.

45. For an even bolder idea on reshaping the Army combat force structure that would make brigade-sized units rather than divisions the key building block, see Douglas A. Macgregor, *Breaking the Phalanx* (Westport, Conn.: Praeger, 1997), pp. 48–93.

Germany, Spain, and Belgium. Finally, the Ace Rapid-Reaction Corps allows a broader set of NATO countries to plan in advance for a joint operation near or beyond the alliance's borders.[46]

Continually debating and reshaping security architectures does little, however, to redress physical gaps in capabilities. In fact, it may even serve as a distraction from such nuts-and-bolts issues. For all this alphabet soup of new initiatives, European capabilities have not changed much. Only the United States is really in a position to deploy large numbers of forces well beyond its national borders and operate them there for an extended period. This is particularly true for regions beyond Europe, but it even applies for places like the Balkans. Rather than focus exclusively on security institutions or wait for them to evolve to the point where a pan-European force could team with the U.S. military for out-of-area operations, the individual European countries—in consultation with each other, but in the end nationally—should focus more on their actual deployable military capabilities. Focusing on national capabilities also makes sense because decisions about using force will probably have to be made at the national level, regardless of the state of European institutions, at least for a time. As Catherine Kelleher puts it, "Military operations require in democratic states a legitimate political mandate before citizen lives will be risked."[47]

Without a major power-projection initiative, political and economic trends now in evidence across the Atlantic may actually exacerbate Europe's current shortcomings in this area. Specifically, NATO Europe on the whole is cutting defense spending but failing to consolidate its major defense industries in the process. As a result, too few marks and francs and pounds and lira are keeping afloat too many production facilities and associated communities. In this context, European countries will continue to lack funds for a major initiative on strategic transport and logistics. What is more, those latter defense priorities will continue to rank lower for them than aerospace and electronics capabilities. Funding the latter defense systems will

---

46. John E. Peters and Howard Deshong, *Out of Area or Out of Reach?: European Military Support for Operations in Southwest Asia* (Santa Monica, Calif.: RAND Corporation, 1995), pp. 97–117.

47. Catherine McArdle Kelleher, *The Future of European Security: An Interim Assessment* (Brookings, 1995), p. 151.

satisfy Europeans' desire for prestige, commercial spin-offs, and foreign sales more than buying roll-on/roll-off ships, tactical transport
trucks, or mobile hospitals.[48]

European governments and militaries should realize that defense of
national territory is no longer the proper focus of their military planning. They should recognize that, in words uttered in 1996 by former
British defense minister Michael Portillo, their forces have become
"hollow," particularly as concerns their deployability to places like the
Persian Gulf.[49] Having the capacity to transport a couple of tens of
thousands of relatively lightly armed soldiers quickly to a trouble spot,
and needing many months to send significant amounts of armor, is
inadequate. Despite spending two-thirds of what the United States
does on defense, European NATO countries have less than 10 percent
of the transportable defense capability for prompt long-range action. It
is largely for that reason that no post–cold war U.S. defense strategy
has assumed significant European contributions in any war plan for
military scenarios in northeast or southwest Asia.[50]

With multiple centers of decisionmaking and no immediate
prospect of realizing a true European Security and Defense Identity
(ESDI), NATO Europe will be hard pressed to make as efficient use of
its defense dollars as the United States. Fashioning integrated forces
out of small building blocks from various countries imposes daunting
demands in the realms of interoperability and battlefield cooperation
that will continue to challenge even an experienced alliance.[51]
Whatever Europe's aggregate capabilities, it would clearly be unwise

48. Canada, though still spending more on defense than all but five NATO
European countries, is left out of this chapter's main discussions simply because it
is not in Europe. Politically, it has a different sense of identity; militarily, its forces
are not based close to those of the European powers, making joint training and
joint transport and logistics operations less convenient than for NATO neighbors.
They are also increasingly focused on peacekeeping missions; together with coastal
patrol, that precludes a significant contribution to traditional air-ground combat
forces. See William S. Cohen, *Report on Allied Contributions to the Common
Defense* (1996), p. II-9.

49. Michael Portillo, "European Security, NATO, and 'Hard Defence,'" speech
at the Royal Institute for International Relations, Brussels, October 23, 1996.

50. See for example, Cohen, *Report of the Quadrennial Defense Review*, p. 24.

51. Paul R. S. Gebhard, "The United States and European Security," Adelphi
Paper 286, International Institute for Strategic Studies (London: Brassey's, 1994),
p. 42.

of the United States to assume that all its allies would agree to participate in a given operation under U.S. or NATO leadership. But all that said, the allies should be able to do much better.

To play a major partner role with U.S. forces, NATO European countries do not need to develop an independent forced-entry capability with large numbers of amphibious ships or air assault units. Nor do they require major initiatives in reconnaissance assets, theater missile defenses (TMD), or several other key support areas. By focusing on strategic transport, deployable logistics, and munitions stocks, they could develop roughly half as much rapidly deployable military capability as the United States for a total of $50 billion over five to seven years. Although large in total amount, that represents an average of well under 10 percent of projected defense budgets. It could be funded within projected resource levels simply by reducing the size of the unnecessarily large European force structures to make smaller numbers of units more usable in theaters where the Western alliance faces its most likely future tests.[52]

As noted, changes in U.S. strategy need not be hostages to any such evolution in European capabilities. The present requirement to be able to decisively and promptly win two overlapping major theater wars in which initial battles are lost and Desert Storm–like reinforcements are required is unnecessary. A somewhat more modest, though still robust, posture would suffice, with or without more European help.

CHANGE OVERSEAS NAVAL PRESENCE

Much of the demand for a large fleet of warships is driven today by the way the Navy conducts its forward presence. For example, eleven active-duty carrier battle groups are needed to maintain an average of not quite three continuous deployments. The reason is principally because the Navy does not wish to keep crews overseas for longer than six months at a stretch and needs to allow them time for classroom training and exercises in home waters. As a consequence, any given crew is usually on deployment for just six months of every twenty or more. Because of the time required to sail back and forth to

52. See Michael O'Hanlon, "Transforming NATO: The Role of European Forces," *Survival*, vol. 39 (Autumn 1997), pp. 10–13.

deployment zones, it is often on station for only about four of those twenty months. That translates into a requirement for five or six ships in the fleet for every one continuous overseas deployment.[53]

Even more battle groups than are now in the force would be needed to maintain nearly three deployments were it not for the homeporting of an aircraft carrier in Japan. The carrier and associated escorts there are on call and therefore "deployed" even when the ships and crews are in port. This arithmetic explains why a fleet with eleven active-duty carrier battle groups can, more or less, maintain continuous deployments in northeast Asia, the Mediterranean Sea, and the Persian Gulf. (Actually, because the ratio is a little worse than 5:1 for a deployment in the Persian Gulf, fifteen carriers are needed to maintain those three continuous deployments.)[54]

"Rotate Crews, Not Ships"

Might naval crews instead be rotated by airlift to ships that remained on forward station for perhaps two years at a stretch (until major overhauls were needed)? Such a policy could allow a given deployment to be maintained with only about 2.5 groups of ships, on average, rather than 5 or more. This concept has been thoroughly explained by William Morgan of the Center for Naval Analyses.[55] Reportedly, the Navy has taken an interest in it for future ships not yet built— although there is no apparent reason why it could not be employed with current ships since flying hundreds or even thousands of sailors from stateside to an overseas port would not be particularly difficult.[56]

Opponents of this approach often argue that it is easy to suggest it from the comfort of a modern Washington office but quite another

53. Ronald O'Rourke, "Naval Forward Deployments and the Size of the Navy," Congressional Research Service, November 13, 1992, pp. 13–23.

54. Les Aspin, *Report on the Bottom-Up Review* (Department of Defense, October 1993), pp. 49–51.

55. William F. Morgan, *Rotate Crews, Not Ships* (Alexandria, Va.: Center for Naval Analyses, June 1994), pp. 1–9.

56. Robert Holzer, "Revolutionary Concept Goes to Navy Chiefs," *Defense News*, August 11–17, 1997, p. 1; and interview with Admiral Jay Johnson, U.S. chief of naval operations, *Defense News*, January 19–25, 1998, p. 22.

thing to make it work in practice. Every ship is different, they argue, and the logistics of moving 6,000 people around the world to man a carrier battle group cannot be understood by an armchair analyst. Admittedly it would be challenging to make this concept work in the real world, but the counterarguments ring hollow in the end. Most Navy ships (including aircraft carriers) are part of general classes of vessels with similar overall features from one ship to the next. Even if individual ships and the machines on them have their idiosyncrasies, it is implausible that today's highly skilled and highly trained sailors cannot learn the peculiarities of two ships rather than just one. Moreover, some key individuals could help smooth the transition by staggering their schedule from the main crew rotation and being available to help a fresh crew learn the ropes of a ship it has not operated before. Given the added airlift capacity that would be purchased in my alternative option for warfighting purposes, the airlift demands would be easily handled. (They could probably be handled by today's fleet, or, if it proved less expensive, this could be contracted out to commercial carriers.) Transporting 6,000 sailors per deployed carrier group could be done with a couple of dozen of the 100 additional military aircraft that would be purchased under this book's defense proposal.

Critics might argue that, even if routine presence could be handled this way, unexpected contingencies might stress the crew-rotation scheme excessively. But the BUR and the QDR determined that such needs were greatest in the case of two simultaneous major theater wars and that those requirements were roughly 20 percent less than the demands of the presence mission. Given the likely availability of land bases for U.S. aircraft in South Korea, Japan, and the Gulf Cooperation Council countries of the Persian Gulf region, even that requirement seems overstated. It is worth noting that even in 1990, when the United States had fifteen aircraft carriers in its fleet, the Gulf War regional commander requested only four.[57]

The various sanctions enforcement and crisis response activities under way since the Gulf War have typically involved no more than

---

57. See Les Aspin, "An Approach to Sizing American Conventional Forces for the Post-Soviet Era," U.S. House Armed Services Committee, Washington, D.C., January 24, 1992, p. 13.

twenty-five to thirty vessels. Even in the cold war 1980s, deployments rarely exceeded forty ships at a time.[58] (In the event of more serious contingencies, maintaining peacetime operational tempo rates would be a moot point.)

Such burdens could be handled by a fleet of about 100 major surface combatants, instead of the QDR's 116. The overall fleet would decline to about 275 ships from its planned level of 330.[59] That would amount to a decline of about 15 percent in the size of the Navy—but the new crew-rotation approach would be about twice as efficient as today's. In fact, it should allow the three normal carrier deployments to be maintained continuously (without today's frequent gaps in coverage).[60]

Even if future needs accelerated greatly for a brief spell, unlikely as that appears, other recourses would be available. Naval presence does have great virtues by comparison with ground forces, but these virtues are greater in some places—like the politically sensitive Persian Gulf—than in places like the ally-rich Mediterranean Sea region or the Korean peninsula region. If carrier presence in one of those latter areas had to be suspended for a period, for example, U.S. tactical fighter capabilities might be temporarily increased in countries such as Italy, Spain, South Korea, or Japan. Granted, not every ally could be counted on to agree to such expanded land presence, and even in places where it is granted there are sometimes restric-

58. Consider a prominent example of where deployments have been maintained recently, the Persian Gulf region. About twelve to fifteen ships have typically been in the Persian Gulf region at any time, including an aircraft carrier and nearly 10,000 sailors. See Charles J. Hanley, "Oil Lubricates 'Special Relationship,'" *Washington Times*, April 16, 1997, p. 10; and "Senators Question Gulf Troops," *Washington Times*, March 31, 1997, p. A10. In early 1998, numbers increased to two carriers and twenty ships overall. See "U.S. Forces in the Gulf," *Washington Times*, February 18, 1998, p. A15.

59. It is worth noting that the quarantine operation in the Persian Gulf was maintained by some fifty to sixty ships during most of the fall of 1990; see Les Aspin and William Dickinson, *Defense for a New Era: Lessons of the Gulf War* (New York: Brassey's, 1992), p. 79.

60. Morgan, *Rotate Crews, Not Ships*, pp. 1–9; John D. Goetke and William F. Morgan, *Review of Surface Ship Presence in Major Deployment Hubs—1976–1982* (Alexandria, Va.: Center for Naval Analyses, 1994), p. 34; and John Robinson, "14 Carriers Needed for Warfighting Requirements, CNO Says," *Defense Daily*, January 28, 1997, p. 129.

tions on the use of the aircraft.[61] But this Air Expeditionary Force concept was already employed with good effect in places like Bahrain, Qatar, and Jordan in 1996.[62] The recent change in the U.S.-Japan Defense Cooperation Guidelines that allows additional Japanese ports and airfields to be made available to U.S. forces if needed also could help with this new approach.[63] In true emergencies, most allies will in the end come through.

The immediate dollars at issue are large, not only in the operating costs of the ships but in procurement expenses for destroyers (and one aircraft carrier). The Navy has already received or contracted for a total of almost seventy Aegis-class vessels with modern radar and fire-control systems. Twenty-seven are cruisers, the remainder are DDG-51 destroyers. The administration plans to buy about fifteen more DDG-51 vessels, at roughly $1 billion apiece. But none of them would be needed according to the proposal put forth here, and the 1998 buy of destroyers could be the last.

In fact, the DDG-51 issue is largely independent of any decision that might be made to reduce the size of the carrier fleet. Even without such a reduction, the Navy could apply the new policy just to surface ships and reduce its planned purchases of DDG-51 destroyers. Destroyers are not particularly needed for escorting carriers across the open ocean. That means that some DDG-51s could remain forward deployed and rely on crew shuttling via airlift even if carriers, with their much larger crews, continued to operate out of U.S. ports. For example, each carrier could have one permanent escort and join its second escort when arriving near forward station.

Whatever approach was taken for aircraft carriers, most destroyer deployments could be conducted through the crew-rotation policy. That means that instead of the fifty-seven Aegis-radar-equipped destroyers now planned, the Navy should be able to get by with roughly thirty to thirty-five. Indeed, even fewer might do, since in many ways CG-47 Ticonderoga class cruisers and DDG-51 Arleigh

61. For example, the Air Expeditionary Force deployed to Bahrain in 1995 had to be reduced in size at the request of the host government; see John Christenson, "Nothing Can Match Carrier Presence," *Defense News*, February 26–March 3, 1996, p. 19.

62. U.S. Air Force, *Air Force Issues Book 1997* (1997), p. 12.

63. See "U.S.-Japan Guidelines for Defense Cooperation," September 23, 1997, sec. V.2.a.

Burke class destroyers have similar functions like air defense and missile attack. But since forty-two have already been authorized, it makes sense to stick with that number.

The implications of this option would clearly be serious for the nation's shipbuilders and raise the issue of whether to reduce the size of the country's shipbuilding base. Specifically, both Bath Iron Works in Maine and Ingalls Shipbuilding in Pascagoula, Mississippi, would suffer considerably under this change in destroyer production.

Today, having six major shipyards to sustain the 330-ship QDR fleet may not be necessary, but it does not appear blatantly inefficient either. In the 1970s and 1980s nine major U.S. yards averaged producing about 1.5 large naval vessels annually; that is roughly the steady-state rate implied by sustaining the BUR/QDR fleet with six yards.[64] However, given my proposal for a fleet of 15 percent smaller size and 20 percent less value (given the large cuts in aircraft carriers), a shipbuilding base of only four or five major yards may make more sense—unless some of the six yards are able to diversify into commercial work or take on more Navy repair tasks. My proposed initiative on sealift could provide a small cushion in the near future, but probably not enough to sustain six yards. Unfortunately, the alternative is to introduce huge inefficiencies into a U.S. defense budget that, to paraphrase former defense secretary Dick Cheney, can no longer afford to be a jobs program.

The Okinawa Marines

This is the right place to mention one other idea, motivated more by alliance politics than by military efficiency, concerning forward presence. Notably, the 20,000 U.S. Marines on Okinawa in Japan, though a capable military force for certain missions, appear now to be causing serious strains in the bilateral alliance. It makes sense to consider ways to bring home all but those manning the 31st Marine Expeditionary Unit (MEU), which conducts routine ocean patrols, as well as those maintaining equipment and staging facilities on the island. Most major U.S. bases and training ranges on Okinawa could be returned to Japan's control in this way. If the decision to make this

64. Ronald O'Rourke, *Navy Major Shipbuilding Programs and Shipbuilders: Options for Congress*, CRS Report 96-785F (September 1996), pp. 41–70.

change was made within the next few years, it could also permit Japan to save the money that would otherwise have been spent replacing the Marine Corps Futenma Air Station—and perhaps devote some of it to other alliance priorities, like theater missile defense. Under the proposal outlined here, that station would no longer be needed. The modest number of flights needed routinely for the 31st MEU could be conducted out of the Air Force's Kadena base; in a crisis or conflict, Naha International Airport could be made available as a staging base for a larger Marine operation in the region if Tokyo supported it.

Mostly because of the Marine presence, U.S. military bases continue to cover 18 percent of Okinawa's territory. That number is due to decline to about 16 percent if and when the Futenma Marine air base is relocated and other changes agreed to in 1996 are instituted. But it will still be down only modestly from the 21 percent figure at the time of the island's reversion to Japan in 1972. Okinawa has as many people as the state of Hawaii on less than one-tenth the land and is densely populated even by Japanese standards. Given the huge benefits that the United States gains from having Navy and Air Force bases in Japan, it is prudent to protect those assets (which require much less land than Marine training facilities) and stop putting the health of the alliance at risk over the less than critical Marine Corps presence.[65]

More than 80 percent of Japanese consider this arrangement unfair, according to two recent polls, and many Okinawans are livid over it. Mike Mochizuki, Satoshi Morimoto, Takuma Takahashi, and I have developed an alternative approach that would keep only 3,000 to 5,000 Marines on Okinawa, in exchange for Japan taking a somewhat greater role in the alliance than the recent U.S.-Japan Defense Cooperation Guidelines will allow and playing at least limited roles in any alliance combat operations. It would also have Japan purchase equipment and storage facilities to keep as much Marine equipment on Okinawa in the future as is there now.

The main concrete downside for U.S. planners would be budgetary: if the Marines were returned stateside, or to a less cash-rich

65. For a fuller description of the relative benefits of the different types of bases the U.S. military possesses in Japan, see Michael O'Hanlon, "Restructuring U.S. Forces and Bases in Japan," in Mike Mochizuki, ed., *Toward a True Alliance: Restructuring U.S.-Japan Security Relations* (Brookings, 1997), pp. 149–78.

country than Japan (like Australia and perhaps also South Korea), the United States would lose the host-nation support payments for these Marines and see their annual cost go up by about $300 million.[66] The assumption of those added annual costs by the U.S. armed forces is included in the option developed in this book.

In a broader sense, however, U.S. policymakers would be even more troubled by the symbolism of retrenching the U.S. military presence in Japan and East Asia more generally. Having declared in the Defense Department's 1995 report on the East Asia-Pacific that U.S. forces in the region would remain roughly 100,000 strong and having reiterated that position in 1996 and 1997, they are unwilling to back away from the figure. Any consideration of redeploying the Marines to a place like Australia, which would preserve the total number of U.S. troops in the region, has apparently been ruled out in light of the continued North Korean military threat and tension between Taiwan and the People's Republic of China. Whether or not counterbalancing military steps could be taken to compensate for the redeployment of the Marines, officials worry about causing even the appearance of a weakening of U.S. commitment to the region or northeast Asian subregion.

As a practical matter, it may be necessary to take stronger additional steps before scaling back the Marine Okinawan presence. But before considering that issue, a word of background is useful. Working within the constraints of Japan's postwar "peace constitution" and the 1960 U.S.-Japan security treaty, Tokyo and Washington have just completed new defense cooperation guidelines. If the Japanese Diet approves implementing legislation, this union between two countries that account for 40 percent of global military spending will be reaffirmed and strengthened.

The most notable revisions to the alliance concern conflicts that do not directly endanger Japan. Under guidelines in place since 1978, Japan has been able to do little more than allow U.S. forces to use bases on its territory. Now, it will also be able to provide those forces nonlethal equipment like fuel and open up other ports and airfields to them as well, permitting, for example, the use of Naha International Airport by U.S. forces in a crisis, as mentioned above. It could also resupply U.S. ships during a crisis and evacuate civil-

66. See Mochizuki, ed., *Toward a True Alliance*, pp. 24–28, 138–43, 149–78.

ians or wounded U.S. soldiers from dangerous situations, provided its own forces stayed out of range of hostilities. In addition, Japanese warships will be authorized to remove mines from the high seas and to help monitor compliance with UN economic sanctions against a given country.

But Japan's military will remain banned from conducting most dangerous missions outside its national territory. Even if U.S. military units were in dire danger, or if the international community authorized the use of force via Security Council resolution, Japan could not put its own armed forces in harm's way. So these new guidelines themselves are unlikely to strike policymakers in Tokyo and especially Washington as a sufficient symbolic and strategic counterweight to withdrawing the Marines from Okinawa.[67]

The best next step would probably be further evolution in Japan's role as a U.S. ally, allowing it to participate in combat during multilateral military operations outside its territory. But so soon after the 1997 guidelines review, that is unlikely to occur in the near future. More likely is that additional U.S. forces of some other kind, either airpower or Navy ships, could be stationed in Japan. With much less land required for such capabilities, and many fewer military personnel involved—perhaps 1,000 to 7,000, depending on the specific type of unit—it would be possible to argue that U.S. forces in Northeast Asia had retained just as much strength as in the past with fewer troops. Most of the Okinawa Marines could be redeployed to Australia or South Korea or returned to the United States. This type of step would make good sense and should be considered by policymakers.[68]

REDUCE NUCLEAR FORCES WHILE ENHANCING SAFETY AND DEFENSES

Although they are becoming less important budgetarily, nuclear and nuclear-related forces still cost some $30 billion to $35 billion a year. That total figure includes Department of Energy (DoE) environmental remediation, intelligence activities, and missile defense

67. See Mike M. Mochizuki and Michael E. O'Hanlon, "Japan as Full-Fledged Ally," *Christian Science Monitor*, November 12, 1997, p. 20.

68. O'Hanlon, "Restructuring U.S. Forces and Bases in Japan," in Mochizuki, ed., *Toward a True Alliance*, pp. 168–73.

research.[69] Some further savings are possible from these accounts, to the tune of about $2.5 billion annually—although those savings would be largely or wholly swallowed up by any decision to deploy a national missile defense that could be reached in the next decade or so.

Over time, though probably not within the budgetary horizon of this study, the United States should also attempt to move down to a posture of only several hundred nuclear warheads. That would allow deterrence against the types of threats, such as chemical or biological attack by a terrorist state, that are most plausible—while keeping at least some linkage between the sizes of arsenals and strategic common sense. The Clinton administration's recent decision to drop planning for a protracted nuclear war with Russia is a step in the right direction on this front. But the current policy of keeping some 3,500 strategic warheads under START II and planning to keep at least 2,000 under a possible START III Treaty is excessively cautious. Indeed, it is cautious to the point of being dangerous. For example, dropping plans for a protracted nuclear war with Russia while keeping large arsenals has reportedly led targeters to broaden their target sets in China.[70] That is an entirely counterproductive type of message to be sending to a country with which we are trying to build good relations, and against which we already have whatever nuclear deterrent benefits the U.S. arsenal may be able to offer. Rather than remain stuck in this paradoxical position, the administration should get moving on further nuclear cuts.

Offensive Nuclear Forces

Under the still-unratified START II Treaty, U.S. and Russian nuclear forces would have to drop from their current numbers of more than 6,000 to between 3,000 and 3,500 deployed long-range warheads. The U.S. warheads would be based on 500 Minuteman III intercontinental ballistic missiles (500 warheads), 71 B-52H and 21 B-2 bombers each carrying about 14 or 15 warheads (1,300 total bombs), and 14 Trident sub-

69. See Stephen I. Schwartz, ed., *Atomic Audit: The Costs and Consequences of U.S. Nuclear Weapons since 1940* (Brookings, 1998).

70. R. Jeffrey Smith, "Clinton Directive Changes Strategy on Nuclear Arms," *Washington Post*, December 7, 1997, p. 1.

marines with 24 missiles each holding 5 warheads (1,680 weapons, for a grand total of just about 3,500).[71]

It makes sense to move to that force posture now, even before Duma ratification of START II, and also to withdraw those remaining tactical nuclear weapons still in service from both overseas bases and operational status. Part of the reason for taking that risk is that, in military terms, it is no risk at all, the destructive capability of even 3,500 warheads being far more than deterrence could truly require in this era. In addition, the conventionally converted B-1 bomber fleet would represent a latent nuclear capability with a capacity to carry about 1,500 more warheads. It could be restored to a dual-purpose role should 3,500 warheads really not seem adequate at some future date. Finally, as the commander of the U.S. Strategic Command, General Eugene Habiger, pointed out in 1996, Russia will be unable to keep forces larger than those allowed in START II beyond 2005 or so in any event.[72] Although unilateral arms cuts are always politically contentious, this approach would follow the spirit of Ronald Reagan's (and the Soviet Union's) decision to respect SALT II limits even though that treaty was not verified and George Bush's 1991 decision to unilaterally eliminate a number of tactical nuclear weapons from the U.S. arsenal while reducing the alert levels of part of the strategic force. Moreover, as noted it would be partially reversible, given the large latent nuclear capability of the B-1 bomber force—although the chances that restoring the arsenal to a greater size would ever seem necessary are particularly low in light of Russia's decaying nuclear weaponry.

In addition to adopting a START II posture right away, the United States could deploy those 3,500 warheads more economically. Specifically, it could equip Trident submarines with the full number of warheads they were designed to carry. Doing so could allow some Trident submarines and Minuteman III ICBMs to be retired and permit immediate termination of production of the D5 missile. Specifically, if all 92 bombers carried on average 15 warheads (1,380 warheads), and if the Minuteman III ICBM force was curtailed to 200 missiles, 10 Trident subs each carrying 8 warheads per missile would round out the

71. "U.S. Strategic Nuclear Forces, End of 1996," *Bulletin of the Atomic Scientists*, January–February 1997, p. 70.

72. "Strategic Command Chief Sees Russian Nuclear Forces Degrading by 2005," *Inside the Pentagon*, August 29, 1996, p. 5.

force at 3,500 warheads. Since Trident submarines remain reliably invulnerable when deployed in the oceans, since at least some missile silos would survive even a complete surprise attack, and since bombers could always be returned to runway alert should future circumstances with the other two legs of the triad change, this approach would not risk putting too many eggs in too few baskets.

DoE's Warhead Efforts

Further savings appear possible in the Department of Energy's nuclear weapons complex—even if one wishes to keep significant latent design capability in that complex under a comprehensive test ban treaty. The two independent design laboratories at Los Alamos, New Mexico, and Livermore, California, have historically been instrumental in pushing innovation in the nation's nuclear capabilities. Whatever the importance of such innovation before, it is probably much less in the post–cold war era. Also, Sandia, the laboratory responsible for everything but the "nuclear guts" of nuclear weapons over the years, has always provided peer review for its technologies internally. The logical approach would be to give Los Alamos, the laboratory that designed five of the seven warheads to be retained in the U.S. arsenal, this mission.[73]

One cannot rule out that problems may develop in the U.S. nuclear arsenal in the future and that the absence of nuclear testing could complicate attempts to redress those problems. Today's highly optimized individual warheads do not allow large margins for error from degradation of key materials or components. Electronic components and certain other parts of the warhead can be individually tested, but compression and deuterium-tritium "boosting" processes in the plutonium pit of a warhead are difficult to mimic. The uncertainties associated with that process will grow when today's highly sensitive plutonium cores need to be rebuilt, particularly since new methods will

73. David Mosher, *Preserving the Nuclear Weapons Stockpile under a Comprehensive Test Ban* (Congressional Budget Office, 1997), p. 65; and "Global Nuclear Stockpiles, 1945–1997," *Bulletin of the Atomic Scientists*, November–December 1997, p. 67. (The latter publication states that six of nine remaining warhead types will be Los Alamos designs; the discrepancy is probably due to whether modifications of a basic design are counted.)

be used for rebuilding them (like casting the metal rather than machining it to the desired shape).

Such worries are probably only serious if one believes that the complete integrity of the single integrated operational plan (SIOP) is needed to ensure deterrence. A somewhat lesser standard seems consistent with the history of nuclear deterrence and even more so with likely future demands upon the U.S. nuclear arsenal.[74] The numbers of different types of warheads in the planned START II arsenal mitigate those concerns because not all seven would be likely to develop problems at once. Moreover, one would not be forced into a posture of attacking cities if the SIOP was degraded. For example, a major attack against the chief fixed targets of a large conventional army might be conducted with about 200 to 300 warheads.[75]

Should serious doubt about the viability of the arsenal ever arise, two or three simple high-reliability weapons could be designed and built without testing—such as a gun-assembled uranium bomb, a "dense pack" plutonium bomb, or even a "levitated pit" plutonium device.[76] Reliable tritium boosting might be sacrificed in the process and a premium paid in terms of the warhead's weight, yield, or safety. But reliable fission bombs and even simple thermonuclear bombs could almost surely be constructed with high confidence without testing.

In this light, two changes are possible. The independent nuclear design capabilities at Lawrence Livermore National Laboratories could be phased out over five to seven years, allowing that laboratory to focus on nonnuclear weapons, verification and intelligence, nondefense research, and the dual-purpose national ignition facility for fusion research. Also, readiness activities at the Nevada Test Site could be ended. Should the Comprehensive Test Ban Treaty someday prove untenable and nuclear tests appear necessary, they could still be conducted there—though it might take several years to prepare to

---

74. McGeorge Bundy, *Danger and Survival* (Vintage Books, 1988), pp. 588–607. There are, admittedly, still some who argue that the details matter, not only for symbolic but also concrete military reasons. An example is the head of Sandia National Laboratories, Paul Robinson. See "Military Can Meet Threat with 2,000 Nukes, But Not Less, Official Says," *Inside the Air Force*, March 28, 1997, p. 12.

75. See Congressional Budget Office, *The START Treaty and Beyond* (October 1991), pp. 22–24.

76. Michael O'Hanlon, *The Bomb's Custodians* (Congressional Budget Office, July 1994), p. 26.

carry one out in the manner employed in the past. Adopting these two policies could save almost $300 million annually—two-thirds of that by consolidating weapons labs, the rest by shutting down the Nevada Test Site. Even with those cuts, DoE's nuclear weapons research budget would remain about three-fourths its cold war average.[77]

Reducing Nuclear Alert Levels

A further set of changes is also desirable at present—namely, reducing the alert levels of some or all U.S. and Russian deployed nuclear weapons. That would respond to the poor state of Russian nuclear forces and the associated risk inherent in its launch-on-warning policy. The Russian posture is especially dangerous in light of the deterioration of its nationwide system of warning devices (at least two of nine early-warning satellites are out of service at present, and two-thirds of the country's modern early-warning radars are inoperable).

To reduce each side's capacity for a disarming first strike and thereby mitigate this danger, some warheads could be physically removed from their missiles. Others could be precluded from rapid launch by pinning open the safety switches on their missiles in a way that would require several hours to reverse. Other types of dealerting would require more research and trial and error to develop in a verifiable way, but that too should be a priority.[78]

Although the cost savings associated with this approach would be minimal, the security benefits could be considerable. However low the odds of a nuclear accident may be at present, they are too high for comfort.

The major counterargument to taking this type of approach is similar to the argument against further cuts in offensive forces (beyond START II ceilings), since reducing alert levels would be tantamount to reducing the size of the strategic arsenal available promptly to national decisionmakers. Some still argue that deterrence requires 2,000 or more warheads, given the way in which the SIOP is determined and their conviction that its approach is time-tested and sound. Others might worry less about being able to rapidly attack a large target set

77. Mosher, *Preserving the Nuclear Weapons Stockpile*, pp. 17–27.
78. Bruce G. Blair, Harold A. Feiveson, and Frank N. von Hippel, "Taking Nuclear Weapons off Hair-Trigger Alert," *Scientific American* (November 1997), pp. 82–89.

but be concerned that further reductions would bring superpower forces down nearer the levels of the medium nuclear powers and jeopardize U.S. nuclear superpower status.

Both these arguments are weak. One can recognize the need to deter attack against the United States or its allies with weapons of mass destruction, and disagree with the viewpoint of those who would eliminate all nuclear weapons, without subscribing to the illogic of cold war targeting requirements.[79] As noted above, possession of several hundred warheads would enable devastating attacks against conventional military targets or other types of key assets related to a country's national power and allow policymakers to avoid targeting cities as their only recourse for employing weapons of mass destruction.

As for superpower status, Russia's rapid descent from the ranks of the world's great powers should underscore how nuclear weapons themselves accomplish little for a state. Moreover, Britain and France have only 200 to 400 nuclear warheads of strategic range (taken here as 5,000 kilometers or more), of which a fraction are on alert at a time, and China has only a few dozen warheads configured with strategic delivery vehicles.[80] In other words, each of the medium nuclear powers has on the order of 100 or fewer strategic warheads on alert at a time, so an American decision to reduce its alert levels would not jeopardize whatever small benefit may accrue to the United States from its advantage in this domain.

National Missile Defenses

Although it is unlikely to affect defense planning immediately, a possible decision to build and deploy defenses in the next decade needs to be kept in mind as one conducts medium-term defense planning and anticipates the coming budget crunch.

79. Eliminating nuclear weapons would, in my view, be unverifiable into the indefinite future and therefore imprudent to attempt by treaty; it would also be imprudent in view of the need to deter not only nuclear but also chemical and biological attacks against U.S. forces or allies. For one discussion of how these concerns relate to a specific regional problem, see O'Hanlon, "Stopping a North Korean Invasion."

80. Shannon Kile and Eric Arnett, "Nuclear Arms Control," in Stockholm International Peace Research Institute, *SIPRI Yearbook 1996* (Oxford: Oxford University Press, 1996), pp. 616–19.

The arguments for and against strategic missile defenses are well known. Arguing in favor, the possibility of a new missile threat to U.S. territorial security arising in the next fifteen years or so, while small, is not zero.[81] At present the United States is defenseless against such an attack. And a small nationwide defense of U.S. territory against an attack by a few or at most several dozen missiles is not inconsistent with stable deterrence among the major nuclear powers. That is partly because U.S. and Russian arsenals remain so large that they could overwhelm any light defense. But it is also because higher-altitude defenses of the type needed to provide nationwide coverage at reasonable cost are vulnerable to fairly simple countermeasures, such as chaff and decoys, which a major power could deploy fairly easily.[82]

On the other hand, in the short term any deployment of missile defenses that interfered with the imperative to improve the safety of Russian nuclear forces would be unwise. Russia may already worry that only several dozens of its warheads would be absolutely sure to survive an all-out U.S. attack. Under those circumstances, it might find deployment of even a light U.S. missile defense reason enough to retain its hair-trigger nuclear posture of launch-on-warning and to rule out any reduction in nuclear alert levels desired for safety reasons.[83] Also, the security of nuclear weapons in a country characterized by poorly trained and compensated troops, as well as widespread mafia-like crime and poor border controls, is worrisome enough that the director of the FBI, Louis Freeh, recently assessed the chances of a nuclear detonation as greater today than during the cold war.[84] The

81. See Bill Gertz, "Report on Missile Threat to U.S. Too Optimistic, Woolsey Charges," *Washington Times*, March 15, 1996, p. 10; and statement of Richard Davis before the U. S. Senate Select Committee on Intelligence, "Foreign Missile Threats: Analytic Soundness of National Intelligence Estimate 95-19," NSIAD-97-53, General Accounting Office, December 4, 1996.

82. It is for such reasons that then Pentagon acquisition chief Paul Kaminski described his vision of any initial U.S. national missile defense as designed to intercept only five to twenty warheads—and even then, only if they were unaccompanied by sophisticated penetration aids. See "Kaminski Urges Caution on NMD Deployment," *Aviation Week and Space Technology*, March 3, 1997, p. 45.

83. Blair, Feiveson, and von Hippel, "Taking Nuclear Weapons off Hair-Trigger Alert," pp. 82–89.

84. See Douglas Farah, "FBI Chief: Russian Mafias Pose Growing Threat to U.S.," *Washington Post*, October 2, 1997, p. A18.

ability of the U.S. government to help redress this situation is limited. But continuing on with DoD's Nunn-Lugar program as well as DoE's program to help control Russian fissile materials is essential.

If deploying national missile defenses reduced Moscow's likelihood of agreeing to dealerting, downsizing, safeguarding, and reforming its military and nuclear forces, or reducing the overall size of its defense budget to expedite economic recovery, it would be unwise—at least at this point in the denuclearization and arms control process. That conclusion is reinforced by the technical immaturity of missile defense technologies, notably evidenced in several failed THAAD interceptor tests in 1997 (see below). But if Russia could be convinced to tolerate light defenses even as offensive forces were scaled back, put on lower alert, and made more secure internally, the deployment of a nationwide missile defense might be desirable. A modest defense—say, six ground-based sites each with a radar complex and twenty-five to fifty interceptors—might be built for $15 billion to $50 billion.[85] But that presupposes that it will someday become technically feasible, which is by no means a foregone conclusion. That raises the matter of theater missile defenses, where research is further along but serious technical problems remain.

Theater Missile Defense

Theater missile defenses against threats like the Iraqi SCUD are important, both in their own right and because advanced TMD systems may have strategic implications someday. For example, the Navy upper-tier defense could provide wide coverage. If coupled with space-based infrared sensors like the surveillance and missile tracking system (SMTS) now under development, it might have strategic potential. And there may be ways to test these systems in conditions

85. The recent RAND and CBO lower-range estimates of $3 billion to $4 billion apply to a single-site defense with multiple radars and satellite links—but nevertheless with questionable ability to defend Alaska and Hawaii even against a very limited attack; see Bill Gertz, "Missile Defense System's Price Tag Drops," *Washington Times*, June 7, 1996, p. 12; Joseph C. Anselmo, "Minuteman 3 Emerges as Viable NMD Option," *Aviation Week and Space Technology*, April 21, 1997, p. 24; and Stanley W. Kandebo, "U.S. Pursues NMD System to Prepare for 'Rogue' Threat," *Aviation Week and Space Technology*, March 3, 1997, p. 44.

simulating those that would be encountered against a strategic reentry vehicle without violating any treaty, especially in light of the recent accord with Russia to limit only the speed and range of a target reentry vehicle but not the characteristics of the interceptor.[86]

However, even fairly simple TMD systems are having major trouble against much simpler and slower threats. Missile defense of any kind is highly challenging technically. In the Gulf War, the Patriot in its then-existing "PAC-2" form performed relatively poorly. There are no absolutely certain confirmations of successful interceptions of incoming SCUD missiles; there is clear evidence of many failures (although the Army and the manufacturer maintain that roughly 50 percent of all SCUDs were successfully intercepted).[87]

There have been improvements to technology since then, widely recognized as promising for the cause of theater missile defense. For example, as a result of initial post–Gulf War modifications as well as the so-called PAC-3 configurations 1 and 2, Patriot radars can now track more objects at once, see them at greater range, pick up fewer false targets from extraneous radar reflections, and exchange information with each other's computers. Interceptors can be based further away from radars, allowing greater coverage area. Starting in 1999, a new missile and more radar upgrades will allow a PAC-3 configuration 3 capability that should for the first time allow the radar to discriminate automatically between a heavy warhead and light decoys or debris and also permit greater maneuverability of the interceptor itself. The system might be able to defend out to several dozen kilometers from the missile's launch point. Such a system should have a good chance of reliably handling the type of threat that Iraq possessed in 1991—particularly if the attacker employs single warheads rather than numerous smaller chemical or biological munitions.[88]

86. See George Lewis and Theodore Postol, "Portrait of a Bad Idea," *Bulletin of the Atomic Scientists*, July–August 1997, pp. 18–25.

87. Robert M. Stein and Theodore A. Postol, "Patriot Experience in the Gulf War," *International Security*, vol. 17 (Summer 1992), pp. 199–240.

88. David Hughes, "Patriot PAC-3 Upgrade Aimed at Multiple Threats," *Aviation Week and Space Technology*, February 24, 1997, pp. 59–61; Lisbeth Gronlund and others, "The Weakest Line of Defense: Intercepting Ballistic Missiles," in Joseph Cirincione and Frank von Hippel, *The Last 15 Minutes* (Washington: Coalition to Reduce Nuclear Dangers, 1996), pp. 57–58.

The next generation of TMD includes the theater high-altitude area defense (THAAD) and the Navy's lower-tier system. Likely to be more effective than Patriot, they will still be considerably limited in range of coverage.

However, these systems are making only limited progress in tests to date. In 1997 Pentagon acquisition chief Paul Kaminski put the Army's THAAD program officials on notice that their program would have to start completing intercept tests successfully or risk a termination of funding.[89] Since then, another test failure has occurred.[90] The Navy's lower-tier program with the Standard Missile 2 Block 4A interceptor, guided by the Aegis radars already deployed on cruisers and destroyers (as well as the interceptors' own terminal homing devices), recently had a successful test. But it was against a nonmaneuvering object with a predictable and known flight path, giving little reason for confidence about the more rigorous tests that are not expected to take place until the turn of the century.[91]

Due to budget constraints, missile defense research in general in the United States relies heavily on specific tests to individual systems as well as computer simulation of system performance. Integrated tests of entire systems are rare and expensive. They also must try to mimic conditions that could be highly variable in practice (depending on the number of incoming targets, their trajectories, and so on). These facts add particular uncertainty to a type of military technology that is in its infancy.[92] Given the huge technical challenges facing any missile system, theater or national in scope, Congress should not impose any specific systems, defense architecture, or timelines for deployment on the Pentagon. Even those who doubt the continued desirability of the ABM Treaty should not throw caution to the wind on deploying systems of uncertain technical

89. "Do or Die?," *Aviation Week and Space Technology*, February 24, 1997, p. 19; and Gronlund and others, "Weakest Line of Defense," pp. 45–60.

90. John Mintz, "Missile Defense System Fails Fourth Test," *Washington Post*, March 7, 1997, p. G1.

91. David Hughes, "Navy Readies Fleet for Anti-Scud Warfare," *Aviation Week and Space Technology*, February 24, 1997, pp. 61–63.

92. William B. Scott, "Mix of Simulation, Flight Testing Troubles BMDO Leaders," *Aviation Week and Space Technology*, February 24, 1997, pp. 64–67; and David Mosher, "The Grand Plans," *IEEE Spectrum* (September 1997), pp. 38–39.

promise and strategic benefit. On the other hand, Congress should be commended—and after the QDR, so should the Clinton administration Pentagon—for ensuring that enough funds are devoted to missile defense research to permit more realistic testing and other needed efforts.

Defense proponents should not be overly fixated on the ballistic missile threat, either. Cruise missile attack may be just as serious a concern, particularly for protecting deployed forces overseas. Defeating that threat would require not only sophisticated computing and communications but fundamentally different types of sensors, since cruise missiles do not produce large rocket plumes and fly at low altitude.[93] Similarly, smuggling of nuclear, chemical, and biological weapons into the country is also a difficult threat. It may be impossible to eliminate, but does appear possible to mitigate through more detection devices like gamma ray cameras at borders and through legal provisions to ensure that the military can respond to a chemical or biological weapon within U.S. borders.[94]

At the end of the day, all of these approaches to protecting U.S. national territory make sense; the only complicated issues relate to deployment of national missile defense. Given the tens of billions of dollars that the United States spends annually hedging against the risk of regional conflicts in places where it has interests that are important but often not vital, the hundreds of millions and few billions of dollars that may be needed for various programs to protect U.S. territory should not be seen as inordinately expensive. On the other hand, the challenges associated with these efforts are immense, and the technologies may not be available in the near- to medium-term future regardless of the wishes and timelines of policymakers.

So much for the alternative force posture. It looks fairly close to the QDR force in broad brush, with identical numbers of divisions and

93. See David A. Fulghum, "Cruise Missile Threat Spurs Pentagon Research," and "Stealth, Cheap Technology Complicate Defense Schemes," *Aviation Week and Space Technology*, July 14, 1997, pp. 44–50.

94. Fred C. Ikle, "Naked to Our Enemies," *Wall Street Journal*, March 10, 1997, p. 18; and Bryan Bender, "Pentagon Found Ill-Prepared for Asymmetric Warfare," *Defense Daily*, October 8, 1997, p. 5.

fighter wings (although each is trimmed somewhat compared with current practice). The Navy fleet would be scaled back about 15 to 30 percent in the realms of aircraft carriers, surface combatants, and submarines. Mobility assets would be enhanced. Nuclear forces would be reduced to START II levels and deployed and maintained more economically, allowing some savings in the short term (though possibly just freeing up money for missile defenses later down the road). With this broad structure in place, it is time to address the question of how to fill it out with equipment.

4
TOPICS IN DEFENSE ACQUISITION

re there ways to keep the U.S. armed forces' equipment stocks reliable and safe, and allow targeted technical innovations, without following the ambitious modernization plans found in the QDR? Do certain planned weapons modernization efforts make more sense than others, given the geopolitical and technological context? By contrast, is a possible military revolution being missed out on?

This chapter addresses these questions. Rather than review all major elements of the acquisition plans of the Department of Defense, it focuses on those areas where good arguments appear to exist for rethinking some of the suggested policies of the Quadrennial Defense Review and where relatively large dollar sums are at issue. (Some questions of acquisition policy are also considered in chapter 3 in sections dealing with strategic lift, naval operations, and nuclear forces.) Before examining these specific issues, it first briefly treats the related question of defense industry consolidation and then offers a perspective on the so-called revolution in military affairs (RMA).

The principal theme of this chapter is as follows. Although weapons modernization remains important, and weapons safety and reliability remain crucial to the U.S. military, the QDR's acquisition plan focuses too much on expensive platforms like combat ships and planes. By taking a more discriminating approach to purchases of tactical aircraft and attack submarines, and reducing the DDG-51 destroyer buy as suggested in chapter 3, the safety and reliability of U.S. military equipment can be ensured at lower cost. A procurement spending level that at present will have to grow to at least $65 billion next decade (from an anticipated 2003 level of $54 billion under the QDR) would only need grow to about $60 billion. After needed investments in lift and pre-positioning were completed, it could drop several billion dollars below that figure.

Modernization would be ensured through a two-tiered approach. First, push hard on the relatively economical and high-payoff improvements in munitions, communications, computers, and sensors that are becoming possible today—and that are sometimes described, albeit in overly grandiose terms, as making possible a "revolution in military affairs." Meanwhile, make more selective and modest acquisitions of new types of major combat systems like fighters and helicopters and combat ships.

However, taken by itself, this generality verges on a platitude. It does little good to argue in the abstract that we have too many fighter programs, for example—as the recent National Defense Panel (NDP) unfortunately contented itself with doing.[1] Only by directly taking on specific "rice bowls" at the Pentagon is one forced to articulate detailed military arguments that can be scrutinized and reacted to by others. That is the first step toward developing the type of firm political consensus actually needed to carry out real and inevitably painful cuts. This chapter attempts to take that first step for a number of high-priority weapons systems.

DEFENSE INDUSTRY CONSOLIDATION

The United States defense industry has consolidated to a remarkable degree in the 1990s, as shown in table 4-1. On average, major defense sectors have one-third fewer companies operating within them than at the end of the cold war. Pentagon acquisition spending, including procurement as well as research, development, testing, and evaluation (RDT&E), has declined from a 1982–91 real average of $140 billion (as expressed in 1998 dollars) to less than $85 billion in 1997 and

1. National Defense Panel, *Transforming Defense: National Security in the 21st Century* (Arlington, Va.: December 1997), p. 49.

TABLE 4-1. U.S. Defense Industry Consolidation[a]

Industrial sector	Percent real change in funding, 1990–98	Number of companies, 1990	Number of companies, 1997
Expendable launch vehicles	−10	6	3
Fixed-wing aircraft	−70	8	4
Munitions	−80	9	9
Rotorcraft	−60	4	3
Satellites	−40	8	5
Strategic missiles	−90	3	2
Surface ships	−75	8	5
Submarines	−50	2	2
Tactical missiles	−60	13	7
Tactical wheeled vehicles	−45	6	4
Torpedoes	−25	3	2
Tactical combat vehicles	−65	3	2

Source: John B. Goodman, Deputy Under Secretary of Defense (Industrial Affairs and Installations), briefing, "Defense Industry Restructuring" (Brookings Institution, June 3, 1997).

a. Assumed 1998 funding based on fiscal year 1998 president's budget. For ships and submarines, only procurement funding is reflected in these data.

1998; even if the desired increases in procurement spending occur, it will remain at only about $100 billion. At that point, the net reduction in outlays will be roughly proportionate with the reduction in typical numbers of major suppliers, suggesting in broad terms that a proper balance is being struck.[2] Moreover, given the increasing complexity of modern weaponry, the same number of inflation-adjusted dollars today typically buys fewer numbers of units than in the past. That may lead the Pentagon to purchase fewer new types of vehicles, planes, and ships in the future—trying to profit from a standard design (like the planned new surface combatant, SC-21, or a common chassis for a number of future ground vehicles). Such trends may mean that consolidation will have an even greater economic logic than budget trends themselves imply.[3]

Whether or not consolidation deserves credit for it, U.S. firms have been improving their labor productivity faster than the (less

2. See Office of the Under Secretary of Defense, Comptroller, *National Defense Budget Estimates for FY 1998* (March 1997), pp. 110–11.

3. See Congressional Budget Office, *Limiting Conventional Arms Exports to the Middle East* (September 1992), pp. 10–13.

FIGURE 4-1. *Productivity in the Defense Electronics Industry,*
1991 and 1995[a]

U.S. dollars

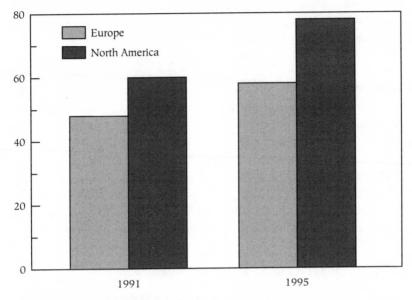

Source: John J. Doudy, "Winners and Losers in the Arms Industry Downturn," *Foreign Policy*, no. 107 (Summer 1997), p. 94.
a. Value added per employee hour.

consolidated) European industry, as shown in figure 4-1.[4] Although it is beyond the scope of this analysis to reach definitive conclusions on the subject, the alternative to consolidation may well be inefficiency, lower productivity, and higher costs.

Concerns about monopolistic tendencies are understandable, particularly when huge mergers like the Lockheed–Martin Marietta, Lockheed-Martin/Northrop-Grumman, and Boeing–McDonnell Doug-

---

4. John J. Dowdy, "Winners and Losers in the Arms Industry Downturn," *Foreign Policy*, no. 107 (Summer 1997), pp. 88–101. It is at least possible that European firms' overall productivity trends (including the effects of both capital and labor) compare more favorably with the United States than labor productivity alone, but that is doubtful.

las consolidations occur.[5] However, it is worth noting for the sake of perspective that monopolies and duopolies have been around a long time in the defense world. There are but one tank producer (General Dynamics Land Systems Division), one builder of aircraft carriers (Newport News Shipbuilding), and two attack submarine producers (Newport News and Electric Boat). Competition is generally desirable, but in some cases it may not be essential—or at least may not require multiple suppliers in a given sector.

What about Pentagon financial support for these consolidations in the form of reimbursing companies for the direct costs of mergers? It is probably a good idea. Because of the "cost-plus" nature of most defense contracting, firms may have only limited incentive (or even disincentives) to get lean and efficient on their own.[6] The government may have little choice but to step in if it wishes to reduce the amount of overcapacity in the industry. Many critiques of mergers and acquisitions in the defense industry seem to ignore the huge costs of an industrial sector staying larger than market conditions warrant.

A "REVOLUTION IN MILITARY AFFAIRS"?

Much of the defense community is intrigued by the notion that a "revolution in military affairs" may be occurring. For proof, one need look no further than the reports of the Quadrennial Defense Review and National Defense Panel, both of which use the phrase repeatedly and favorably.[7] This RMA school holds that advances in precision munitions, real-time data dissemination, and other modern technologies can help transform the nature of future war and with it the size and structure of the U.S. military. The RMA hypothesis is central in any discussion of acquisition policy. If major breakthroughs in warfare are imminent and require large new investments to realize, that would suggest the need for a major change in Pentagon spending priorities that at present are focused largely on force structure and readiness.

5. See, for example, Lawrence J. Korb, "Merger Mania," *Brookings Review*, vol. 14 (Summer 1996), pp. 22–25.

6. I am indebted to Kenneth Flamm for counsel on this subject.

7. See William S. Cohen, *Report of the Quadrennial Defense Review* (Department of Defense, May 1997), pp. iv–v, 39–51; and National Defense Panel, *Transforming Defense*, pp. 5–8.

Reasons to Be Skeptical of the RMA

The visionaries may not be right—or at least may be exaggerating their case. Rapid technical innovation in the military sphere has, in fact, actually been going on for a considerable time. It is not clear what is now different about its pace or relative significance.

The twentieth century has witnessed constant technical break-throughs with enormous consequences for strategy and warfare. Even since the advent of blitzkrieg warfare and carrier-based airpower, the following technologies or capabilities have been incorporated into modern military forces: radar, helicopters, infrared sensors for guidance and targeting, laser-guided bombs, laser rangefinders, high-performance jet engines, stealth technology, autonomous and accurate missiles, reconnaissance satellites, and the modern high-speed computer (not to mention thermonuclear weapons and intercontinental ballistic missiles).[8]

Are the munitions, sensors, and integrated communications systems now being developed or produced of even greater significance than the above technologies? That seems doubtful. They should be incorporated into the force, but their overall effectiveness should probably be seen as evolutionary rather than revolutionary.[9]

Not too much should be made of the semantics, however. Perhaps one can say that there is an RMA under way—but if so, it is because, as Israeli defense expert Jonathan Shimshoni writes, "An entrepreneurial military organization must be in a state of permanent revolution; change must be its constant condition."[10] In other words, the current information-led defense modernization wave can be seen as the RMA that is following a multitude of previous RMAs—including most recently those catalyzed by high-performance jet aircraft, infrared and laser detection and targeting sensors, satellites, stealth

8. The nineteenth century may have been equally impressive, characterized as it was by the advent of the mass national army, industrial production of armaments, railroad logistics, the rifled barrel, and the invention of the machine gun. See John Keegan, *A History of Warfare* (Vintage Books, 1993), pp. 301–66.

9. For a similar perspective, see Williamson Murray, "Thinking about Revolutions in Military Affairs," *Joint Forces Quarterly* (Summer 1997), pp. 69–76.

10. Jonathan Shimshoni, "Technology, Military Advantage, and World War I: A Case for Military Entrepreneurship," in Steven E. Miller, Sean M. Lynn-Jones, and Stephen Van Evera, *Military Strategy and the Origins of the First World War*, rev. ed. (Princeton University Press, 1991), p. 160.

technology, and cruise missiles. This transformation, with its emphasis on system networking and integration, may be different than the others, but that does not mean it is necessarily the most important.

A good deal of the enthusiasm about the RMA seems derived from the impression that computer technology is completely revolutionizing life, the global economy, and therefore inevitably warfare. But as Stephen Biddle argues, the effects of the computer revolution on society as a whole may be considerably more limited than appearances suggest. Although good for Silicon Valley, it may not do as much for the economy—or the military.[11]

For those who would make a great deal of Desert Storm and assume its lessons can be quickly built upon in the future, it is worth remembering several caveats about the performance of high-technology equipment in that war. Infrared, electro-optical, and laser systems were all seriously degraded in performance by weather, dust, and smoke. Even high-resolution radars on aircraft such as the F-15E had difficulty distinguishing tanks from trucks at tactical distances. And advanced munitions did not, in this case, have to operate against Iraqi countermeasures.[12]

That list of problems highlights a serious limitation upon a "revolution" driven by innovations in computer and communications: however fast and efficiently an advanced electronic grid can transmit data within a military force, those data will only be useful when reliable and accurate. Computers and communications cannot compensate for many of the limitations on sensors that are likely to continue to inhibit performance in the foreseeable future. Without good data, the old adage about computers applies: garbage in, garbage out.

While a good deal of optimism exists about coming breakthroughs in sensors, much of it is unwarranted. A good example is the recent Air Force claim that, "In the first quarter of the 21st century you will be able to find, fix or track, and target—in near real-time—anything of consequence that moves upon or is located on the face of the Earth."[13] That may be true for easily recognizable military objects like

11. See Paul Krugman, "The Paper Bag Revolution," *New York Times Magazine*, September 28, 1997, pp. 52–53.

12. See General Accounting Office, *Operation Desert Storm: Evaluation of the Air Campaign*, NSIAD-97-134 (June 1997), pp. 32–36.

13. See Statement of General Ronald R. Fogleman, Chief of Staff, U.S. Air Force, before the House National Security Committee, May 22, 1997.

tanks and ships (provided they can be distinguished from allied and civilian assets). It is unlikely to be true, particularly at standoff distances, for small arms, for enemy soldiers interspersed among a background population, for mortars and antitank and antiaircraft missiles hidden inside trucks or caves, and for properly secured weapons of mass destruction. The Air Force is probably more wrong than right on this one.

Even if some of these weapons or military systems prove to be detectable and trackable, they will often not be identifiable as enemy capabilities until the enemy chooses to initiate hostilities by taking the first shot or shots—particularly in infantry and counterinsurgency warfare. And even if they are picked up by reconnaissance assets when they are fired, their ability to "shoot and scoot" may make it impossible for standoff weaponry to reach them before they can hide in buildings, crowds of people, or complex terrain. To give a sobering example of the limits of current technology, only about two of the sixty mortars that were fired in Mogadishu during the U.S. operation there in 1992–94 were spotted and targeted quickly enough to be destroyed.[14]

The sensor challenge extends not only to satellites, spy planes, and unmanned aerial vehicles but also to homing "brilliant" munitions. A number of new types of ordnance, intended to pick up the infrared or acoustic signature of enemy equipment and then attack it with small shaped charges, are now being acquired. They include the sensor-fuzed weapon (SFW), brilliant antitank submunition (BAT), and others. But regardless of how well they are doing on test ranges, none have been challenged by a determined adversary to date. In addition to building battlefield flares to confuse infrared sensors, a foe might design countermeasures to blare out noise to jam acoustical sensors and generate large amounts of dust and smoke to cloud the vision of sensors on homing munitions. Attacks might be waged during heavy fog or cloud cover.

Even if the United States devises successful counter-countermeasures, autonomous homing munitions of the type now envisioned may only be useful while an adversary is in the attack phase and its forces can be clearly distinguished from allied units and civilian

14. "U.S. Military Operations in Somalia," Hearings before the U.S. Senate Committee on Armed Services, Senate Hearing 103-846, May 12, 1994, p. 41.

assets. That is an important phase of many wars, to be sure. The Desert Storm plus Desert Shield plus Bosnia framework advocated in chapter 3 is based partly on the assumption that modern technologies are improving the lethality and responsiveness of U.S. military forces enough that they can take better advantage of that phase. But that hinges on reacting quickly to any assault. Doing so is a challenging task at both political and military levels. It is much more likely to be accomplished when a solid forward-defense strategy is in place than in a region where the United States does not expect war and may not be bound by a firm treaty commitment to defend a friendly country (but may ultimately choose to get involved anyway). If the United States does not counterattack during a narrow time window when enemy forces are relatively exposed and isolated, it may have a great deal more trouble employing standoff munitions with autonomous homing devices against enemy forces. Technology will remain important, and improved technologies will help across a wide variety of tactical settings, but they may not fundamentally change the nature of battle the way RMA proponents insist.

How to Modernize and Innovate

To be dismissive of the RMA concept as now widely employed would be to throw the baby out with the bathwater. The U.S. military victory in Desert Storm depended heavily on advanced technology, and a technological advantage is not maintained by standing still. Clearly, further invention and innovation are needed. The questions have to do with philosophy and pace of adaptation, not with the need to adapt.[15]

In this light, it is worth remembering that the U.S. post–World War II military establishment brought capabilities like stealth, precision-guided munitions, command and control aircraft, advanced reconnaissance techniques, and sophisticated sensors into its organization without requiring the formal announcement of any revolution. What was needed were financial and human resources to develop and then purchase new technologies, as well as the tactical and operational flexibility on the part of commanders to use them effectively.

15. For a concurring view, see Thomas G. Mahnken and Barry D. Watts, "What the Gulf War Can (and Cannot) Tell Us about the Future of Warfare," *International Security*, vol. 22 (Fall 1997), pp. 160–62.

All these features of cold war U.S. defense policymaking are being retained in the post–cold war era. U.S. military RDT&E spending of about $35 billion a year remains three times the size of the rest of NATO's spending combined, roughly the size of either Russia's or China's entire defense budget by some estimates and considerably higher than the U.S. RTDT&E average of the 1950s, 1960s, or 1970s.[16] It will remain high by those decades' standards even at its scheduled 2002 real level of $31 billion, and will still be more than three-fourths the 1980s average. Three-fourths of that scheduled decline of $4 billion is due to the anticipated completion of advanced development of the F-22, F/A-18 E/F, V-22, and a new attack submarine, so it should not be viewed as a lessening of the country's basic commitment to military research and innovation.[17] Much of the remaining reduction should be achievable by downsizing a DoD laboratory and weapons testing infrastructure of eighty-six facilities that has been largely spared the scrutiny of the base closure process so far but appears ripe for consolidation.[18] Defense research and development (R&D) receives more resources than all other federally funded R&D in the country combined (see table 4-2); even under President Clinton's ambitious new plan for nondefense R&D, unveiled as part of the 1999 budget request, defense R&D will remain about half of the country's total federally funded research effort.

Those large R&D resource levels have been supporting important recent innovations. They include the development of unmanned aerial vehicles (see table 4-3), production of Joint STARS reconnaissance assets, improvements in advanced munitions, further deployment of certain advanced technologies like LANTIRN infrared navigation and targeting systems for attack aircraft, and addition of the Longbow

16. International Institute for Strategic Studies, *The Military Balance 1996/97* (Oxford: Oxford University Press, 1996), p. 40; and Office of the Under Secretary of Defense (Comptroller), *National Defense Budget Estimates for FY 1998*, pp. 110–11.

17. Congressional Budget Office, *Reducing the Deficit: Spending and Revenue Options* (March 1997), pp. 26–27; Lane Pierrot, *A Look at Tomorrow's Tactical Air Forces* (Congressional Budget Office, January 1997), p. 24; and Marine Corps Briefing Materials on V-22 Osprey, June 13, 1997.

18. "Pentagon Considers Restarting 'Vision 21' Lab Consolidation Effort," *Inside the Pentagon*, August 7, 1997, p. 1; and Mark Walsh, "Congressional Hurdles Cloud Pentagon Lab Closure Plan," *Defense News*, June 9–15, 1997, p. 52.

TABLE 4-2. *Federally Funded Research and Development*[a]
Billions of dollars unless otherwise indicated

Category	1997, estimate	1998, president's budget proposal	2002, projected	Percent real change, 1997–2002
Defense (military)[b]	37.5	36.8	35.1	−17.7
Health and Human Services	12.9	13.2	13.4	−8.7
NASA	9.3	9.6	9.3	−11.9
Energy	6.1	7.3	5.9	−15.5
Defense	2.8	3.7	2.6	−19.0
Nondefense	3.4	3.6	3.3	−12.7
National Science Foundation	2.4	2.5	2.6	−7.5
Agriculture	1.5	1.5	1.5	−14.9
Commerce	1.0	1.1	1.3	9.1
Interior	0.6	0.6	0.6	−7.6
Transportation	0.7	0.7	0.7	−7.4
Environmental Protection Agency	0.5	0.5	0.6	7.6
All other	1.2	1.2	1.2	−11.3
Total	73.7	75.0	72.1	−14.0
Defense R&D	40.2	40.5	37.6	−17.8
Nondefense R&D	33.5	34.6	34.5	−9.4

Source: Intersociety Working Group, *Research and Development FY 1998* (Washington: American Association for the Advancement of Science, 1997), pp. 61, 120.

a. Figures may not add to totals due to rounding.

b. The main components of the Defense Department's research and development budget request for 1998 are as follows (in billions of dollars): basic research, 1.2; applied research, 2.8; advanced technology development, 3.4; demonstration and validation, 5.6; engineering and manufacturing development, 8.5; management support, 3.1; operational systems development, 11.3; other, 0.8.

radar to Apache helicopters to give them all-weather capability with Hellfire antiarmor missiles.

This approach of "rapid incrementalism"—adding new capabilities onto existing forces when possible, while also trying to streamline and subtract obsolete capabilities—is the right way to go in the future as well. One does not need to rebuild the military from scratch. A useful historical analogy to bear in mind is that the development of blitzkrieg by Germany between the world wars depended primarily on joining the low-cost radio to forces made up of planes and tanks, which would have been purchased in large numbers in any event, new doctrine or not.

TABLE 4-3. Current and Planned U.S. Unmanned Aerial Vehicles[a]

Characteristics	Pioneer	Hunter	Outrider	Tier II, Predator	Tier II+, Global Hawk	Tier III–, Dark Star
Operating altitude (km)	4.6	4.6	1.5	4.6	15.2–19.8	15.2
Maximum endurance (hours)	5	11.6	4 (+ reserve)	35	38 (20 at 5,556 km)	12 (8 at 926 km)
Radius of action (km)	185	267	200	740	5,556	926
Maximum speed (km/hr)	204	196	222	204–215	639	556
Cruise speed (km/hr)	120	165	167	120–130	639	556
Sensors[b]	EO or IR	EO and IR	EO and IR[c]	EO, IR, and SAR	EO, IR, and SAR	EO or SAR
System composition (AVs)	5	8	4	4	*	*
Number of systems	9	7	6	12	5[d]	4[d]

Source: Defense Airborne Reconnaissance Office, *UAV Annual Report FY 1997* (Department of Defense, November 1997).

* Not yet determined.

a. All systems also include various ground stations that are not mentioned here.

b. EO = electro-optical; IR = infrared; SAR = synthetic aperture radar.

c. SAR growth potential.

d. Planned.

Another reason for incrementalism is that nothing about improved computers and communications will soon end the need for battlefield protection for troops. Armor in some form is likely to be needed well into the future. The Pentagon's goals of achieving "focused logistics," "full dimension protection," and "precision engagement" will therefore not end the era of large weaponry, large support forces, or other trappings of traditional armies.[19] Taken together, technical trends suggest that the rapid evolution that has characterized warfare for at least the entire twentieth century is simply continuing.

DoD would be wise to keep pushing. But it is less important to explicitly try to usher in a revolution in military affairs than to continue what has brought us frequent mini-revolutions throughout recent decades: robust support for R&D in general.

The need for balance and patience in achieving military innovation is underscored by the lack of specificity in the acquisition proposals of most RMA enthusiasts. For example, the National Defense Panel simply called for added spending of $5 billion to $10 billion a year in areas such as intelligence, space, urban warfare, joint experimentation, and information operations. But these are such broad categories that most major weapon systems can be said to involve or advance at least one of them—and if any new system did not, it would probably at least contain one of the following additional attributes that the NDP report also considers to be important for future forces: automation, mobility, stealth, speed, range, and precision strike.[20]

In addition to the vagueness of these new investment categories or criteria, creating a funding wedge, as the NDP suggested, invites gamesmanship at the Pentagon. One can bet that the respective services would be smart enough bureaucratically to emphasize the "revolutionary" or "transforming" nature of various features on their future surface combatants, fighters, helicopters, and other preferred new systems. In fact, the only hardware they might be unable to associate with the NDP's criteria for future procurement is the unglamorous logistics, transport, engineering, and other support capabilities that the services have often underfunded in the past. The NDP report risks encouraging them to do so in the future. It also

---

19. See William S. Cohen, *Annual Report to the President and the Congress* (Department of Defense, April 1997), pp. 69–73, 143–55.

20. National Defense Panel, *Transforming Defense*, pp. 45, 59.

risks a rush to buy technologies for their own sake—raising the risk that DoD may acquire systems before they can be reliably evaluated or wisely chosen.[21]

As noted above, the Pentagon has large amounts of funds that it still devotes to acquisition. If increased to the extent that traditional analysis suggests, as outlined in chapter 2, the funds should be adequate to serve sound modernization purposes. It is beyond this study's scope to suggest exactly which modest-sized joint and innovative programs should be funded more and which traditional service-specific programs funded less. The NDP's case for focusing on systems like sensor suites that might be deployed on small aerial vehicles and aid U.S. forces greatly in urban combat is convincing.[22] But if done correctly, "jointness"—or more effective cooperation between the military services—should simply become an increasing part of normal DoD investment and should save the Pentagon money in general. No new special funding wedge should be needed—especially since introducing such a wedge would simply invite semantic games and other shenanigans as the services tried to make their preferred programs qualify for the "slush fund."

The situation is slightly different with regard to training. The NDP calls for new facilities to allow joint experimentation and practice, and its recommendations are both specific and convincing.[23] Accordingly, my alternative option adds $500 million a year—admittedly a rough figure, but a reasonable one—to fund the recommended initiative. Corresponding to 3 to 5 percent of what the services currently spend on training annually, it would allow several brigades, air squadrons, Marine Expeditionary Units, and other comparably sized units from each service to undertake a vigorous set of efforts to

21. As Fred Kagan and Stephen Biddle point out, not all attempts to revolutionize military technology and operations are successful, and care must be taken to avoid "the wrong RMA." See Stephen Biddle, "Assessing Theories of Future Warfare," paper prepared for the 1997 Annual Meeting of the American Political Science Association, pp. 34–41.

22. National Defense Panel, *Transforming Defense*, pp. 13–15, 43, 68–74; see also Mark Hanna, "Task Force XXI: The Army's Digital Experiment," *Strategic Forum*, no. 119 (National Defense University, July 1997); and Alan Vick and others, *Preparing the U.S. Air Force for Military Operations Other than War* (Santa Monica, Calif.: RAND Corporation, 1997), pp. 63–74.

23. National Defense Panel, *Transforming Defense*, pp. 68–73.

develop new concepts of operations.[24] Once proven, they could be integrated into the normal training regimens of today's units as the NDP's dream of making all training more joint gradually became reality. But it is not clear that at that point, more funds would be needed above and beyond those that support today's already vigorous training programs; in any case, it is too soon to know.

With these general considerations in mind, it is time to turn to specific systems.

TACTICAL COMBAT AIRCRAFT

All U.S. tactical combat aircraft in operation today, including even the F-117 Stealth fighter, were already developed by the end of the 1970s. Major procurement of most of those systems, including the Navy F-14 Tomcat, Navy/Marine Corps F/A-18 C/D Hornet, and Air Force F-15 Eagle, F-16 Falcon, and the F-117 as well, continued through the 1980s. Given that modern fighter aircraft have generally not been operated longer than about twenty years per platform, and that production runs typically must last about a decade (or more) to avoid excessively costly years, the United States will soon need to buy large numbers of planes again.

Current U.S. fighter modernization plans call for three planes to be acquired over the next ten to fifteen years: the Air Force's F-22 Raptor, primarily to replace its F-15C air-to-air fighters; the Navy's F/A-18 E/F Super Hornet, to replace some of its earlier versions of the F/A-18 and the F-14; and the multiservice Joint Strike Fighter (JSF)—a plane desired by all services except the Army to replace Marine Corps AV-8B Harriers, Air Force F-16 Falcons, and Navy F/A-18s. (See appendix tables 4A-1 and 4A-2.) That leaves undecided the successors for the F-15E Strike Eagle and F-117; derivatives of either the F-22 or JSF are likely to be candidates.[25]

Aging U.S. aircraft will require major refurbishment or replacement

---

24. The military spends about $20 billion a year on operations and deployments. See William J. Perry, *Annual Report to the President and the Congress* (February 1995), p. 42.

25. Pierrot, *A Look at Tomorrow's Tactical Air Forces*, p. 5; and Department of the Navy, *1997 Posture Statement* (1997), p. VIII-12.

if U.S. air combat missions are to remain safe for their pilots and dependable for warfighting commanders. Also, it is indisputable that today's airframes generally reflect fairly old technology. But several factors cast into doubt the need for large numbers of new platforms: the slow pace of modernization in other countries; the paucity of modernization funds in "rogue" states and the abysmal capabilities of their current air forces as evidenced by Desert Storm; as well as the ability to modernize U.S. capabilities largely through improved sensors, air-to-air and air-to-ground missiles, and communications (without replacing the platforms themselves). Some new tactical combat airframes make sense—largely to establish air superiority and spearhead strike forces—but not the number envisioned by the QDR.

F-22 Raptor

If built as expected, the F-22 will be the world's preeminent fighter aircraft at least in the first decades of the twenty-first century. It will also possess some air-to-ground attack capability, featuring two 1,000-pound Joint Direct Attack Munition (JDAM) bombs. That capability was added during the program's development in light of the demise of the Soviet threat and the correspondingly reduced need for exclusive focus on the air-to-air mission. (Development of the F-22, then known as the advanced technology fighter, was begun during the peak cold war years of the early 1980s.) Its attack capability is to be of somewhat limited benefit, however, given the lack of precision ordnance like laser-guided bombs or Maverick missiles.[26]

Under the 1997 Quadrennial Defense Review, the planned buy of F-22 fighter aircraft was reduced from 438—itself a reduction from the initial target of 750 planes and 1991 adjusted goal of 648—to 339. The planned maximum annual production rate was reduced from 48 to 36 aircraft.[27] Production of the plane is to begin in 1999 and be completed shortly after 2010 (unless a ground-attack version of the aircraft is then being built to replace the F-15E or F-117).[28] The plane

26. Congressional Budget Office, *Reducing the Deficit: Spending and Revenue Options* (March 1997), pp. 39–40.

27. Cohen, *Report of the Quadrennial Defense Review*, p. 45.

28. United States Air Force, *Air Force Issues Book 1997* (1997), pp. 60–61.

is to be built by Lockheed-Martin, in conjunction with Boeing; engines will be built by Pratt and Whitney.[29]

Most air wings contain 72 primary aircraft. Their pipeline of repair and training airframes is about 25 percent as large, and attrition reserves drive up the total number further. All together, that 339 buy translates into about 3 fighter wing equivalents, assuming the usual wing structure.

The expected R&D costs for the program are about $23 billion, expressed in 1998 dollars. Roughly $20 billion has been authorized through the fiscal year 1998 budget. Significant procurement costs are due to begin in 1999. Even adjusted for the results of the QDR, they will still reach a sustained annual level of at least $2.5 billion throughout most of the next decade. Although DoD claimed, before the QDR, that the plane would cost only $90 million per copy on a unit basis (including production of some standard spares), CBO estimates that the unit procurement cost could reach about $110 million. The General Accounting Office as well as some internal Pentagon cost evaluation groups also believe the Pentagon's current numbers optimistic, thinking that costs could increase 20 percent or so above those now anticipated.[30] That is hardly surprising in light of the F-22's new capabilities, which include stealth designed into the shape of its airframe to give it low-observable qualities reportedly comparable to those of the B-2,[31] engines whose direction of thrust can be varied relative to the plane itself, supersonic speed even without use of afterburners, and advanced integrated avionics.[32] The plane's builder, Lockheed-Martin, hopes to achieve the lower price through many of the same "lean manufacturing" practices that have helped it lower F-16 unit production costs this decade by more than one-third even as production volumes have declined.[33] But its ability to do so is in considerable doubt.

29. Bert H. Cooper, "F-22 Aircraft Program," CRS Issue Brief 87111, December 30, 1996, pp. 1–3.

30. General Accounting Office, *Tactical Aircraft: Restructuring of the Air Force F-22 Fighter Program*, NSIAD-97-156 (June 1997), pp. 1–7; and Pierrot, *A Look at Tomorrow's Tactical Air Forces*, p. 24.

31. Bill Sweetman, "The Progress of the F-22 Fighter Program," *Jane's International Defence Review*, Quarterly Report, no. 1 (1997), p. 17.

32. Pierrot, *A Look at Tomorrow's Tactical Air Forces*, pp. 5–7.

33. William B. Scott, "Lockheed Martin Reconstructs TAS Unit as 'Fighter Enterprise,'" *Aviation Week and Space Technology*, July 28, 1997, pp. 64–66.

What future threat could possibly justify such an expensive air-craft? China, for example, has spent the better part of the 1990s sim-ply acquiring half a wing of Su-27s. That aircraft approaches F-15 caliber in some realms but not in others—and according to U.S. gov-ernment assessments, is not as capable as the F-15 overall. The Office of Naval Intelligence (ONI) apparently expects that China will need until 2005 just to complete a one-wing capability of Su-27s. Moreover, China will still be depending on Russia for aircraft upkeep at that time, and will presumably still have limited training capabilities, pilots without combat experience, and an air force that remains part of the army rather than an independent service. ONI also assumes that China will not be producing anything much bet-ter than the F-10, now in development, in the year 2015. A new Russian fighter is expected to be operational by then and to be a good candidate for exports to countries like China. But whether a Russian economy unable to even properly feed its current soldiers and keep their equipment operational will be able both to design and to produce world-class aircraft—with advanced engines and aerodynamics and sophisticated electronics and sensors—appears open to considerable doubt.[34]

Since investments have already been made in the F-22's technolo-gies and are being counted on for the JSF as well, it makes sense to purchase some of the aircraft despite the lack of a plausible threat in the medium-term. But 339 seems far more than needed. Operationally and militarily speaking, how might a smaller number be logically determined? A good guide is provided by the Gulf War. There, some

34. For number of Chinese Su-27, see International Institute for Strategic Studies, *Military Balance 1996/97*, p. 181. For a comparison of the Su-27 with the F-15, see United States Air Force, *F-22 Raptor . . .* (1997), p. 9, together with Office of Naval Intelligence, *Worldwide Challenges to Naval Strike Warfare* (1996), p. 35. For projections of China's future acquisitions, see the latter report, pp. 29, 35. For the quality and nature of its air forces, see Kenneth W. Allen, Glenn Krumel, and Jonathan D. Pollack, *China's Air Force Enters the 21st Century* (Santa Monica, Calif.: RAND Corporation, 1995), pp. xiii–xxi. For the degree to which Russia relied on a highly inefficient communist economic system to produce high-quality equip-ment in the past—a system it no longer has recourse to—see Clifford Gaddy, *The Price of the Past: Russia's Struggle with the Legacy of a Militarized Economy* (Brookings, 1996); and "Officers Reject Yeltsin's Military Plan," *Washington Times*, August 6, 1997, p. 13.

120 U.S. F-15Cs participated, flying a sortie a day on average but two sorties each on the first day of the conflict. Their missions were to sweep the skies of Iraqi aircraft, defend U.S. attack aircraft against any incursions by Iraqi fighters, and protect U.S. ground forces and large refueling and control aircraft further removed from the front lines of battle. They were aided by about half as many Saudi F-15Cs, which flew somewhat less than half as many missions.[35] The largest group of aircraft used in a single mission was the 24 F-15s and F-14s that conducted an independent "fighter sweep" on the first night of the war, waiting for any Iraqi aircraft that might have been launched in response to the initial coalition air raid. More commonly, F-15s operated in groups of two or four, either on smaller fighter sweeps or as escorts for attack packages.[36]

In the Gulf, F-15Cs were also aided by about 100 Navy F-14s and 90 Navy F/A-18s. However, the Navy has its own modernization programs for these aircraft (see below). One might further argue that F-15Cs could have been aided in the air defense or air superiority missions by the 250 U.S. F-16 aircraft in the Persian Gulf theater, even though they generally were not assisted. But the Air Force does not need F-22s for that mission since it will retain F-16s for now and then replace them with JSF aircraft.[37] Also, the F-22 has a similar weapons-carrying capacity to the F-15C (which generally flew with four Sidewinder and four Sparrow missiles during the Gulf War),[38] and is supposed to exceed its capabilities by about 50 percent in sortie rate per day, cruise speed, and reliability (that is, mean number of flights between major maintenance). That allows for an increased tempo of activity as a cushion against unexpected needs.[39] The intro-

---

35. Department of Defense, *Conduct of the Persian Gulf War* (April 1992), pp. T-56–T-59; and Eliot A. Cohen, *Gulf War Air Power Survey*, vol. 5: *Statistical Compendium and Chronology* (Government Printing Office, 1993), p. 335.

36. Eliot A. Cohen, *Gulf War Air Power Survey*, vol. 4: *Weapons, Tactics, and Training* (Government Printing Office, 1993), pp. 162–70, 192–202, 213.

37. Thomas A. Keaney and Eliot A. Cohen, *Gulf War Air Power Survey Summary Report* (Government Printing Office, 1993), pp. 197–99.

38. Cohen, *Gulf War Air Power Survey*, vol. 4, p. 107; and Timothy M. Laur and Steven L. Llanso, *Encyclopedia of Modern U.S. Military Weapons* (New York: Berkley Books, 1995), p. 96.

39. United States Air Force, *F-22 Raptor . . .* , pp. 11, 16; and United States Air Force, *Air Force Issues Book 1997*, pp. 60–61.

duction of such improved capabilities on next-generation aircraft has led some observers, including former Air Force chief of staff Ronald Fogleman, to suggest that combat air squadrons and wings might be 10 percent to 20 percent smaller at some future date; the F-22s appear good enough that we can begin the streamlining process with this particular aircraft.

An alternative approach to the QDR would be to accept more risk, on the premise that neither Iraq nor Iran nor North Korea will be able to modernize their air forces appreciably over the next decade or so. In that light, having a single theater-war capability in the F-22 force could well be adequate, with a country like China providing a militarily more plausible threat than one of the small "rogue" states. By that one-war logic, and by reference to Desert Storm, purchasing roughly 150 F-22s makes the best sense. That number of planes is large enough to provide a hedge against ongoing improvements in the world's air-to-air missile market and in Chinese capabilities, without spending inordinate sums of money. Those planes could be configured into two smaller wings of about sixty planes each, consistent with General Fogleman's long-term vision.

By producing the aircraft on the same ramp-up path as now planned by the Air Force and then closing down the production line, unit costs might be held to nearly the currently anticipated level. Alternatively, the production line could be built on a smaller scale, with the intention of building only twelve to fifteen airplanes per year. That approach would lead to somewhat higher unit costs but facilitate consideration of a ground-attack variant of the F-22 to replace the F-15E and F-117 after 2010.[40] Purchasing more existing F-15Cs or F-15Es for the remaining 200 aircraft—at about $50 million apiece—would keep the Air Force fighter fleet young and safe while saving nearly $15 billion.[41]

Also important, although much less expensive than fighter modernization, is improvement in U.S. short-range dogfight missiles. The

40. See Greg Caires, "ACC Chief Sees F-22 as Possible F-15E, F-117 Replacement," *Defense Daily*, September 9, 1997, p. 398.

41. See Pierrot, *A Look at Tomorrow's Tactical Air Forces*, p. 3; "McDonnell Douglas Unveils Effort to Sell USAF New Wing of F-15Es," *Inside the Pentagon*, June 5, 1997, pp. 3–4; and "Money: New Production-Cost Target for 339 F-22s Is $43 Billion," *Defense Daily*, July 8, 1997, p. 33.

United States does not expect to wind up in many dogfights, given its
AWACS capabilities and sophisticated fighter-based radars and longer-
range air-to-air missiles. But the possibility cannot be dismissed. The
Russian AA-11 and Israeli Python 4 are considered to lead the world in
this capability, above the U.S. AIM-9M Sidewinder in capability. The
latter is planned for replacement. If there are any troubles with the suc-
cessor program, it would make sense to purchase the Python 4.[42]

F/A-18 E/F

The Navy has virtually completed development and begun production of an expanded, enhanced, and stealthier version of its existing
F/A-18 Hornet aircraft. It is to combine the roles of fighter and attack
aircraft—just as today's A through D models can, but with greater
payload and range for the latter mission without sacrificing agility for
the former.

The F/A-18 E/F is far cheaper to develop than the F-22—even
allowing for recently disclosed problems with its wing that lead to
destabilizing air turbulence during certain dogfight maneuvers and
that may require a redesign of the wings.[43] But it will also represent a
less significant improvement in technology. It will continue, for example, to carry weapons externally (where their radar signatures are
high), employ avionics similar to those on current models, and utilize
fairly traditional engine technology that will do little more than keep
the larger F/A-18 E/F as fast and maneuverable as A/B/C/D models.
The Navy's view is that this aircraft will be adequate for its purposes,
obviating any potential need for a variant of the F-22 aircraft for
itself.[44] It will have considerably greater weapons payload than the
current Hornet, a somewhat smaller radar cross section that may
reduce the range at which it is readily detectable by nearly half, the
ability to return to a carrier with more unused ordnance than current

42. Office of Naval Intelligence, *Worldwide Challenges to Naval Strike Warfare*,
p. 20.
43. See George C. Wilson, "Wing Flaw Dogs Navy F/A-18E/F," *Defense News*,
December 1–7, 1997, p. 3; and Elaine M. Grossman, "Navy Papers Cited Severity
of F/A-18E/F 'Wing Drop' Prior to Milestone Approval," *Inside the Pentagon*,
December 4, 1997, p. 1.
44. George C. Wilson, "Aircraft Dogfight Awaits Cohen at Pentagon," *Defense
News*, January 27–February 2, 1997, p. 4.

planes are permitted to land with, and other upgrades.[45] As a result, it is expected to cost significantly more than the A through D versions—at least $62 million a copy, expressed in constant 1998 dollars, in contrast to roughly $45 million for its predecessors. The Navy expects the F/A-18 E/F to maintain the same technological edge over the Russian Su-35 that today's Hornet has over the Su-27 and to be at least the equal of next-generation European aircraft through 2015 and possibly beyond.[46]

The QDR reduced the anticipated buy of E/F "Super Hornets" from 1,000 to 548, but it reserved the right to restore the planned buy to as many as 785 aircraft should problems develop in the JSF program as the year 2010 approaches. More concretely, the QDR reduced total expected procurement of this system by 24 between 1998 and 2002, and then reduced the maximum planned annual production rate from 60 to 48 for the years 2002 and beyond.[47] The envisioned purchase of about 550 Super Hornets would be roughly adequate to support two squadrons of twenty aircraft within each of eleven air wings, assuming current Navy force structure.

Enhancing the Navy's power projection capabilities and stand-off ranges makes sense. The Persian Gulf has the potential to be inhospitable to carriers in future battle, given increased Iranian acquisition of mines, submarines, and surface-to-surface antiship missiles, as well as other trends. Potential battle with China over Taiwan, for example to help the Taiwanese repulse an attempted Chinese amphibious assault, could be done more safely from greater distances. The A through D Hornets have combat radiuses, depending on weapons loadings and other issues, of 500 to 1,000 kilometers; the E/F is intended to have 40 percent greater range, allowing it to patrol the Taiwan Straits while carriers remain several hundred kilometers east of the island.[48] Also, with the E/F R&D program more than 80 percent complete, and scheduled to cost only about $1 billion more, it makes sense to get something out of this technology effort.

45. Edward H. Phillips, "F/A-18E/F Meets Flight Test Goals," *Aviation Week and Space Technology*, January 20, 1997, p. 58.

46. Office of Naval Intelligence, *Worldwide Challenges to Naval Strike Warfare*, pp. 18, 35; Pierrot, *A Look at Tomorrow's Tactical Air Forces*, pp. 3–8; and Department of the Navy, *1997 Posture Statement*, p. VIII-10.

47. Cohen, *Report of the Quadrennial Defense Review*, p. 45.

48. Laur and Llanso, *Encyclopedia of Modern U.S. Military Weapons*, p. 94.

However, these types of extremely difficult scenarios are unlikely to develop in more than one place at a time. Also, the F/A-18 C aircraft remains superior to China's top-line Su-27 aircraft and to Iraq's MiG-29, according to U.S. Navy Intelligence (without even taking into account pilot training, aircraft maintenance, airborne control, and other areas of considerable U.S. advantage).[49]

In light of these considerations, purchasing roughly 300 E/F aircraft makes sense. For the QDR's eleven-air wing force, that translates into approximately one Super Hornet squadron per wing. In a serious crisis or conflict in which long-range naval airpower proved important, the country could probably swap airplanes if needed. It could double up the Super Hornets on the four to five carriers that might be responding to the highly demanding situation (that is the number of carriers assumed appropriate for regional crises under the BUR and QDR). The remaining need for aircraft could be satisfied by further production of C/D Hornets, saving nearly $5 billion over the next decade by purchasing 250 aircraft at considerably lower cost.

Better yet would be to reduce the carrier and carrier air wing force structure by four, as advocated in chapter 3. In that case, 300 to 350 E/F airframes would be enough for two squadrons on each carrier, and most C/D models could be retired. Total savings in aircraft procurement would exceed $10 billion.

Munitions

More mundane, and less expensive, than fancy fighter aircraft are the munitions that those aircraft deliver. The armed forces would benefit from additional purchases above those now planned.

Some recent critiques of precision munitions, while correct to emphasize that rigorous bomb-damage assessment was not possible in Desert Storm and that the performance of current munitions can be badly degraded by weather or smoke, sometimes went too far. Their point that precision weapons were more expensive than other ordnance dropped in the Gulf War, for example, is true but largely irrelevant. Even these expensive munitions, with a total value for those expended in Desert Storm of less than $1 billion, were far less expen-

49. Office of Naval Intelligence, *Worldwide Challenges to Naval Strike Warfare*, pp. 34–35.

sive than platforms employing them (conservatively worth $50 billion on the U.S. side alone). Also, "tank plinking" and other precise attacks were invaluable for destroying Iraqi weapons in the front-line forces in a way that unguided ordnance could not begin to rival.[50]

Although calculations based on unclassified data are impossible to make, a greater need for precision weaponry is apparent in the services' recent statements to Congress. For example, in 1996 Air Force Chief of Staff Ronald Fogleman stated that $100 million in added funding for precision munitions would be helpful that year. The benchmark that Gulf War precision munitions cost somewhat less than $1 billion is useful. It suggests that one might add $100 million a year to the planned budget through 2010, given the possibility of two conflicts and the need to purchase a new generation of precision munitions as well.[51]

ARMY HELICOPTERS

For all the attention to the three new fixed-wing combat planes now being acquired, the fact that two major armed helicopter modernization programs are also under way is often ignored.

Specifically, the Army is modifying its existing AH-64 Apache attack helicopters, adding a sophisticated millimeter-wave radar to provide all-weather capability with the Hellfire antiarmor weapon (that is also capable against air defense radars and other fixed targets). The Army plans to modify more than 700 Apaches. It also is slowly developing a next-generation scout and light-attack helicopter known as the RAH-66 or Comanche. It intends to build nearly 1,300 of those Comanches eventually, with 700 for the light-attack role and 592 for the scout or reconnaissance role. Those two systems will be complemented by the aging Cobra and Kiowa fleets in the near term, but eventually will constitute the entire Army combat helicopter capability.[52]

50. See, for example, General Accounting Office, *Operation Desert Storm: Evaluation of the Air Campaign*, NSIAD-97-134 (June 1997), pp. 17–41, 181.

51. Heather J. Eurich, "Service Chiefs Split with White House on Modernization Requests," *Defense Daily*, March 14, 1996, p. 392; and Zalmay Khalilzad and David Ochmanek, "An Affordable Two-War Strategy," *Wall Street Journal*, March 13, 1997, p. A14.

52. Frances Lussier, *An Analysis of U.S. Army Helicopter Programs* (Congressional Budget Office, 1995), pp. xv, 22.

Consistent with the philosophy embodied in the above F-22 and F/A-18 E/F options, a reduced buy of Comanches makes sense. But how many? Two basic factors would allow roughly a halving of the current program size. To begin with, in the Gulf War Apaches operated without scout helicopters, calling into doubt the basic need for the Comanche for that latter purpose. Improvements in UAVs and other sensors suggest that heavy attack helicopters may often be able to function independently in the future.[53] A sensible option could therefore eliminate the 592 Comanches intended for the scout mission. Consistent with the reduction of 1.5 division equivalents in the active Army suggested here, the remaining buy of Comanche light-attack aircraft could be further reduced by 10 percent, as could the number of Apaches slated for Longbow upgrades.

This approach would not save the nearly $4 billion still needed to develop the Comanche, but it would save the cost of buying 650 of the helicopters at $20 million apiece, as well as the cost of refurbishing 80 Apaches at $9 million per aircraft. That savings of nearly $15 billion would be realized after 2005, however. Also, it would be more than half offset by the need to improve the Army's utility helicopter fleet. As demonstrated by Frances Lussier of the Congressional Budget Office in a detailed study, that fleet is aging considerably. It is becoming less reliable and more dangerous to those who operate it. The Army should amend this growing problem by purchasing more Black Hawk UH-60 aircraft for troop, weapon, and supply transport. On the whole, this integrated approach would still wind up saving the Pentagon about $5 billion relative to the administration's plan. But even more important, it would redress an imbalance in the Army's aviation priorities, focusing less (though still quite considerably) on technical advances and more on basic reliability and safety.[54]

THE V-22 OSPREY

Having survived a tortured funding history that saw it ended by one secretary of defense, kept alive by the Congress, and then restored to

53. CBO, *Reducing the Deficit*, pp. 53–54.
54. Lussier, *An Analysis of U.S. Army Helicopter Programs*, pp. xv, 17–19, 40.

life by the next administration and reasonably well protected by the QDR, the V-22 Osprey tilt-rotor program is now winding down the R&D phase and entering the production phase. Specifically, 1998 marks the last year that funding for the nearly $7 billion research, development, test, and evaluation program will exceed $500 million; thereafter, annual levels will drop below $300 million (1999), $150 million (2000), and $100 million (2001 and beyond).[55] The current plan is to purchase 360 of the aircraft for troop and light-equipment transport for the Marine Corps, as well as roughly another 100 divided between special forces and the Navy. Ten have been authorized so far, in the 1997 and 1998 defense budgets.

The plane holds promise for certain types of missions, particularly given its speed compared with other transport aircraft capable of operating from amphibious ships or other restricted spaces (all of which are helicopters). With a top speed of about 300 miles per hour (roughly 500 kilometers per hour), it is roughly twice as fast as most helicopters. With a normal mission radius of more than 200 miles (more than 300 kilometers), it also has greater range than most current helicopters (though much less of an advantage than the Marine Corps routinely claims when comparing it against one particular shorter-range system).[56] It also profits from improved systems designed to survive in a nuclear, chemical, or biological environment; reduce the aircraft's infrared and acoustic signatures; and absorb heavier fire than existing helicopters can take before catastrophic failure. Given that its R&D phase has been nearly completed, it makes sense to acquire at least a small fleet of the aircraft.

Caveats about the system also abound. It is incapable of carrying heavy vehicles, meaning that CH-53 helicopters with all their vulnerabilities would still need to be part of any amphibious assault fleet even if V-22s are purchased in large numbers.[57] Its reduced vulnerability to enemy action has often been overstated and is notably better only in regard to small-arms fire (not surface-to-air missiles). Indeed, the well-known 1990 V-22 study by the Institute for Defense Analyses on the topic estimated that a V-22-equipped force would suffer only

55. Marine Corps briefing on V-22 Osprey, June 13, 1997.
56. Ibid.
57. CBO, *Reducing the Deficit*, p. 37.

5 to 25 percent fewer losses than the other options analyzed.[58] As a consequence, its major advantages over traditional rotary aircraft are most relevant to highly time-sensitive or longer-range operations and not as much to large-scale amphibious assault. Its technology remains promising yet untested under realistic circumstances; its operating costs may therefore prove higher than expected.[59] And its production costs are considerably greater than for helicopters with equivalent capability—specifically, about twice as great as for the CH-53E, which has a greater payload and an ability to carry heavy equipment the V-22 cannot.[60]

In light of these various concerns, it would appear most sensible to buy a smaller number of V-22s than desired by the Marine Corps (and stick with the planned numbers of Navy and special forces planes). Converting the anticipated production run of 360 Marine aircraft down to about 150 would allow enough planes to transport about a battalion equivalent unit, since the V-22 can carry about twenty troops (or fewer when also carrying equipment). Perhaps more significant, it would provide sufficient aircraft to use on routine Marine expeditionary unit patrols that are often the nation's front-line reaction force for operations such as embassy evacuation. It would also provide aircraft for search and rescue and for antiterrorist or counterproliferation missions. A single Marine Expeditionary Unit force, of which two to three are deployed at a time, might include some 30 transport helicopters today.[61] Replacing those with V-22s in routine operations would be possible as long as the Osprey fleet numbered roughly 100 or more. Using CBO assumptions of an Osprey unit production cost of about $55 mil-

---

58. Comments of Assistant Secretary of Defense David S. C. Chu at a special hearing before a subcommittee of the Committee on Appropriations of the United States Senate, July 19, 1990, Senate Hearing 101-934, 101 Cong. 2 sess. (Government Printing Office, 1990), p. 51; and L. Dean Simmons, "Assessment of Alternatives for the V-22 Assault Aircraft Program," Institute for Defense Analyses, June 1990, Senate Hearing 101-934, p. 17.

59. Chu, Senate Hearing 101-934, p. 47.

60. CBO, *Reducing the Deficit*, pp. 36–37; and Chu, Senate Hearing 101-934, p. 47.

61. Vincent C. Thomas Jr., ed., *The Almanac of Seapower 1994* (Arlington, Va.: Navy League of the United States, 1994), pp. 144–48.

lion, and $30 million for the helicopter equivalent, some $5 billion would be saved over the next decade.[62]

ATTACK SUBMARINES

About $1 billion must still be spent to complete the new attack submarine R&D program. Even though the need for the submarine could have been called into question before (and a fleet of Seawolf submarines and Los Angeles class subs deployed instead), that remaining amount of funds is small enough to make it cost-effective to purchase the new submarine. Its enhanced shallow-water and special-forces capabilities will be of benefit for the scenarios of greatest concern in the first decades of the twenty-first century.[63]

The attack submarine fleet was to number forty-five to fifty-five vessels under the BUR and is to number fifty ships under the QDR. But the QDR report acknowledges that the SSN (sub-surface vessel, nuclear-powered) peacetime presence mission still needs reevaluation.[64]

The SSN mission in need of greatest revision has to do with monitoring the movements of Russian nuclear-weapons submarines. That mission was always of dubious strategic benefit since it threatened Moscow's survivable deterrent and thereby added to pressures to launch on warning in a crisis. With such a launch-on-warning policy perhaps a greater danger today, given the weaknesses in Russia's early warning networks and general deterioration of its forces, the downsides of putting its submarines routinely at risk have been magnified.[65] The mission should be ended. In fact, in the fall of 1997 the Navy announced that it would end such patrols under the Arctic; it should streamline or end the mission in other places too.[66]

The BUR report stated that a two-war strategy required more than 30 SSNs but less than forty-five (it was no more specific than that, and neither was the QDR). The BUR argued that the requirement for

62. CBO, *Reducing the Deficit*, p. 36.

63. Ibid., pp. 26–27.

64. Cohen, *Report of the Quadrennial Defense Review*, p. 47.

65. Bruce G. Blair, *Global Zero Alert for Nuclear Forces* (Brookings, 1995), pp. 43–72.

66. Patrick J. Sloyan, "Arctic Submarine Patrols Being Phased Out," *Philadelphia Inquirer*, November 17, 1997.

forty-five to fifty-five subs in the fleet arose not from warfighting requirements but the peacetime presence mission.[67] Accepting those Pentagon numbers, but dropping the SSBN trailing mission from the SSN repertoire, a reasonable total fleet size might be thirty-five ships. That number can be calculated as follows.

Begin with a total assumed fleet size, given current missions, of about fifty submarines (the midpoint in the BUR's range). Assume that to maintain normal rotation policies, three attack subs are assumed to be dedicated to each deployed Russian SSBN. At the time of the BUR, the United States probably anticipated that Russia would keep up to fifteen ballistic-missile submarines under the START II Treaty, and that it might be able to maintain up to four at sea at a time.[68] That translates into twelve U.S. attack submarines with the mission of trailing the Russian SSBNs. Subtracting them from the fleet leaves thirty-eight total ships. But since the United States probably also planned against the possibility of a surge in Russian SSBN deployments, some additional American subs might have been assigned this mission as well, leading to my recommendation that an elimination of the tracking mission should lead to a reduction in the U.S. SSN fleet down to thirty-five submarines. This number would allow at least ten attack submarines to be deployed at a time, or several in each key maritime region of the world.

Cutting the submarine fleet at a time when mines and enemy submarines pose serious concerns to U.S. warplanners may seem unwise. But it should be noted that shallow-water antisubmarine warfare (ASW) is conducted largely by sonar arrays trailed by surface ships and backed up by aircraft; submarines are not needed in high numbers.[69]

67. Les Aspin, *Report on the Bottom-Up Review* (Department of Defense, October 1993), pp. 56–57.

68. See International Institute for Strategic Studies, *The Military Balance 1993–1994* (London: Brassey's, 1993), p. 231; and Congressional Budget Office, *The START Treaty and Beyond* (October 1991), p. 154. In 1993, although it seemed reasonable to assume a smaller Russian SSBN fleet, it could not safely be assumed that Russia would permanently reduce the deployment rates of its remaining SSBNs.

69. See Tom Stefanick, *Strategic Antisubmarine Warfare and Naval Strategy* (Lexington, Mass.: Lexington Books, 1987), pp. 35–37, 64–65, 166–73. Although actually finding a submerged diesel submarine can be difficult, effectively quarantining it—and denying it the ability to impede other naval operations—generally is not. It takes time, but can be done, especially if the goal is to protect carriers and other vessels operating in deep water. That is because diesel submarines, even of

Reducing the attack submarine fleet from fifty to thirty-five vessels would save at least $20 billion in procurement costs over the next two decades. It would also do much to invalidate the notion of keeping two independent submarine yards. That conclusion is reinforced by comments of the Navy's top budget planner, Vice Admiral Don Pilling, who argues that even with the current force posture, "We don't need two guys building submarines." The Navy estimates the inefficiency of keeping two production facilities today at about 3 percent of the total new-attack submarine program cost (roughly $2 billion); the inefficiency percentage would increase at a smaller fleet size.[70]

It may make most sense to allocate future work to Electric Boat in Groton, Connecticut. That would keep two independent facilities, Electric Boat and Newport News Shipbuilding, familiar working with nuclear-capable ships—partly as a hedge against any future monopolistic tendencies in one place or the other, partly as a hedge against a future attack against one of the facilities. Hypothetically, two submarine yards might be sustained if most submarine repair work was shifted from government facilities to the losing yard. But that requires a detailed industrial-base analysis beyond the scope of this study.

B-2s and Arsenal Ships

Two other weapons systems of specific interest in the late 1990s time frame are the Stealth bomber and a new type of ship optimized to carry large numbers of missiles with smart munitions that would be interconnected to various communications systems to gain targeting data. The argument is not particularly strong for procuring more than the 21 B-2s already authorized, however, and is interesting though not yet compelling for arsenal ships.

---

the air-independent propulsion variety, cannot move far or fast without depleting their batteries and needing to snorkel. Some risk to amphibious ships, pre-positioning ships, and other vessels that need to reach shore might remain, though mines would probably pose a greater threat than enemy submarines. See W. J. Holland, "Battling Battery Boats," *Proceedings* (June 1997), pp. 30–33; and Director of Expeditionary Warfare, Office of the Chief of Naval Operations, *U.S. Navy Mine Warfare Plan*, 2d ed. (U.S. Navy, February 1994), pp. 24–29, 57–60.

70. See "Splitting Contracting of Sub to Raise Cost by $2 Billion, But Will Save Jobs," *Washington Times*, September 15, 1997, p. 4.

The main argument for buying more B-2s is to be able to search for, detect, attack, and destroy enemy armor while it is exposed early in a future aggressive operation—yet not readily attackable by any assets based nearby. Since a given bomber could carry about 1,000 submunitions, a lethality rate of 5 percent could allow a force of twenty bombers to destroy a heavy armored division in a single pass—assuming good targeting information and submunitions capable of distinguishing real targets from decoys and then penetrating their armor. Allowing for multiple sorties and a larger force, as postulated in the following calculation, it is at least mathematically possible that in certain scenarios such a B-2 force could be very effective.

This scenario is more likely to be relevant to an as-yet unspecified locale than to northeast and southwest Asia, where U.S. tactical aircraft have ready access to land bases in the region. Among the most plausible scenarios might be a Russian attack on Ukraine, the Baltics, or the Caucasus. These are hypothetical military conflicts that I would advocate the United States avoid, given the risks of escalation and the difficulty of consolidating any initial gains that standoff attack might achieve. But some defense planners clearly feel differently.[71]

Regardless of one's views on the need to plan for such scenarios, one needs to ask about other practical matters. One has to do with speed. To be effective, the B-2 would generally need to be used early in a conflict, when enemy armor was isolated from allied and civilian vehicles and could therefore be attacked with homing munitions. (At least for the foreseeable future, those munitions will be incapable of distinguishing between different types of vehicles, so using them in an urban environment or a battlefield where different countries' forces are interspersed will be impractical.) If the United States failed to react quickly against aggression, a larger B-2 fleet would do it little good. And for hypothetical wars in many places, a prompt reaction would be politically difficult to generate out of the American system of government. Even if the United States ultimately decided to oppose a major aggression in a place like central or south Asia, for example, it might need time to reach that determination politically—by which point the window of opportunity to

71. See, for example, National Defense Panel, *Transforming Defense*, pp. 13, 73.

respond with B-2s against exposed and isolated enemy armor could have passed.

The stronger counterarguments against further B-2 production are military. Specifically, even for scenarios such as those mentioned above, there are likely to be cheaper ways of delivering comparable ordnance. If effective homing munitions can truly be developed, as B-2 proponents assume, other means of delivering them would probably exist that were safer and cheaper than using a penetrating bomber. For example, enemy forces could be spotted and targeted by less valuable aircraft or spacecraft, then attacked with long-range cruise missiles carrying smart munitions. Given the remarkable progress in developing new types of reconnaissance assets—ranging from improved spy satellites to unmanned aerial vehicles to Joint STARS aircraft and perhaps someday to F-22s—it seems wisest to separate the tasks of targeting and delivering munitions and have separate platforms carry out the two roles.

Even taking into account the fact that B-2s are reusable and cruise missiles are not, the cost calculations would seem to come out decidedly to the latter's advantage. A B-2 can carry about thirty-six times as much ordnance as a 2,000-km range Tomahawk cruise missile. But even when bought at reasonable production rates, it costs nearly 1,000 times as much as a Tomahawk (the latter costs just over $1 million apiece).[72] Even if the acute attack phase of a future war lasted ten days or so, and each B-2 could make half a dozen flights (as it might be able to if enough munitions, fuel, and spare parts were pre-positioned at places like Diego Garcia, Guam, or Okinawa),[73] it would in effect be carrying 200 times as much as a Tomahawk but still cost almost 1,000 times as much. Moreover, the upkeep costs of the Tomahawk are quite modest, whereas the operating and maintenance costs of the B-2 are at least $20 million per plane per year. That would add another 50 percent or so to the expected bomber lifetime price tag. So instead of being just five times as expensive a way to carry

72. David Mosher, "Options for Enhancing the Bomber Force," Congressional Budget Office, 1995, pp. 24–25, 39; and Elizabeth Heeter and Steven Kosiak, "FY 1998 Defense Authorization and Appropriation Acts: Impact on Defense Programs and Industry," Center for Strategic and Budgetary Assessments, Washington, November 19, 1997, p. 8.

73. Mosher, "Options for Enhancing the Bomber Force," pp. 89–95.

ordnance as cruise missiles, the B-2 bomber fleet could be seven or eight times as expensive.[74]

The conclusion is that B-2s would become economical only if at least several high-intensity wars were fought over their service lives or if enemy forces were highly exposed in their assault phase for considerably longer than one week.[75] The B-2 would also be desirable if the cruise-missile option proved unworkable—perhaps because reliable communications between detection devices and cruise-missile carriers could not be assured.

The argument for the B-2 could also be weaker than suggested above, notably if it proved more vulnerable than now anticipated. Indeed, though details are classified, the B-2 has had considerable trouble in retaining its low observability during sustained testing and operations, apparently maintaining a mission-capable rate less than half the Air Force goal of nearly 80 percent.[76]

As for arsenal ships, the calculus is more appealing since the cost of the vessels is much less than for B-2s. Also, their planned reliance on other systems for targeting data is more logical and consistent with likely future trends in military communications systems. Since modern cruisers and destroyers, of which there would be roughly seventy even under the streamlined fleet advocated here, can typically carry around one hundred Tomahawk cruise missiles each, an arsenal ship could carry as much firepower as five traditional surface combatants or roughly twenty-five bombers. Today, those bombers and surface combatants are bought and paid for, but they will not last forever. At some point, the decision of whether to replace those current capabilities partly with systems like arsenal ships will have to be faced.

An arsenal ship would have the further advantage of being deployable in the waters surrounding a tense theater indefinitely, allowing

74. "CBO Estimates on B-2 Production Costs," *Inside the Pentagon*, May 29, 1997, p. 4; and David Mosher with Raymond Hall, "Options for Enhancing the Bomber Force," CBO Paper, Congressional Budget Office, July 1995, p. 71.

75. For further explanation of why the assault phase is so different from other stages of battle, see Christopher Bowie and others, *The New Calculus* (Santa Monica, Calif.: RAND Corporation, 1993), p. 54.

76. Tony Capaccio, "Air Force Admits B-2 Maintenance Limits Overseas Deployments," *Defense Week*, August 11, 1997, p. 1; and William B. Scott, "Follow-On B-2 Flight Testing Planned," *Aviation Week and Space Technology*, June 30, 1997, p. 48.

even more rapid response than bombers based in the United States. Assuming the rather modest destruction rates of roughly one armored vehicle per missile, consistent with assumptions from a 1993 RAND study for homing munitions, up to a dozen arsenal ships might usefully be stationed near a potential conflict zone.[77] If deployed in those numbers, it could prove a large money saver when compared with likely alternatives. The major caveat to pushing this program is the uncertain potential of smart submunitions when encountering countermeasures and other real-world challenges.[78] A second caveat concerns the possible vulnerability of arsenal ships to attack. Depending on how these technologies progess, arsenal ships may or may not become wise purchases in several years.

With this alternative modernization philosophy and the force structure of chapter 3 now in hand, the final remaining issue concerns the "readiness" and well-being of the individual American soldier, sailor, airman, or Marine. It is to that issue that the final chapter of this study turns its attention.

77. See Bowie and others, *New Calculus*, pp. 54–55.
78. Laur and Llanso, *Encyclopedia of Modern U.S. Military Weapons*, pp. 422–36.

TABLE 4A-1. *Major Pentagon Acquisition Programs, as of June 30, 1997[a]*

Costs in billions of 1998 dollars

Weapon system	Total program cost	Quantity
Army		
Abrams upgrade	7.2	1,060
AFATDS (Advanced Field Artillery Tactical Data System)	1.2	5,299
ASAS (All Source Analysis System)	0.8	28
ATACMS/APAM (Army Tactical Missile System)	2.7	2,465
ATACMS/BAT (Army Tactical Missile System/Brilliant Anti-Armor Submunition)	4.8	21,677
ATIRCM (Advanced Threat Infrared Countermeasures System)/CMWS (Common Missile Warning System)	2.4	2,698
Black Hawk (UH-60L)	3.6	547
Bradley FVS (Fighting Vehicle System) upgrade	5.0	1,602
FAAD C²I (Forward Area Air Defense Command Control Intelligence) BKII/III/IV	1.1	15
Comanche (RAH-66)	8.1	6
Crusader (AFAS/FARV)	2.5	
CSSCS (Combat Service Support Control System)	0.3	1,241
FMTV (Family of Medium Tactical Vehicles)	12.2	85,488
Javelin	3.9	29,015
JSTARS GSM (Ground Station Module)	1.6	161
Longbow Apache	7.9	985
Longbow Hellfire	0.06	13,003
MCS (Maneuver Control System)	1.2	3,156
SADARM (Sense and Destroy Armor)	2.5	73,778
SINCGARS (Single Channel Ground and Airborne Radio System)	4.4	258,896
SMART-T (Secure Mobile Anti-jam Reliable Tactical Terminal)	0.6	313
Subtotal	74.1	. . .
Navy		
AAAV (Advanced Amphibious Assault Vehicle)	0.9	n.a.
AIM-9X	2.6	10,049
AN/SQQ-89	5.4	91
AOE 6 (fast combat support ship—*Supply* class)	3.0	4
AV-8B remanufacture	2.2	72
CEC (Cooperative Engagement Capability)	2.9	206
CVN-74/75	8.8	2
CVN-76	4.5	1
CVN-77	4.7	1

TABLE 4A-1 *(continued)*

Weapon system	Total program cost	Quantity
DDG 51 (destroyer)	57.8	57
E-2C reproduction (early warning and control aircraft, Hawkeye)	3.4	112
F/A-18E/F Hornet Strike Fighter	65.0	1,000
JSOW (Joint Stand-Off Weapon)	8.0	23,800
LHD 1 (amphibious assault ship—USS *Wasp*)	10.0	7
LPD 17 (LX) (amphibious assault ship)	9.5	12
MHC 51 (minehunter, coastal-*Osprey* class)	1.9	12
MIDS-LVT (Multifunctional Information Distribution System–Low Volume Terminal)	1.6	1,607
NESP (Navy EHF Satcom Program)	2.3	399
NSSN (New Attack Submarine)	51.3	30
SH-60R (LAMPS BLKII)	4.9	188
SSN (attack submarine) 21/AN/BSY-2	15.2	3
STD MSL 2 (BLKS I-IV)	11.8	11,658
Strategic sealift	5.9	19
T-45TS	6.4	189
Tomahawk TBIP	1.6	1,253
Trident II MSL	33.0	462
UHF follow-on	2.1	9
USMC H-1 upgrade	2.9	284
V-22 *Osprey*	38.2	523
Subtotal	367.8	. . .
Air Force		
ABL (Airborne Laser)	5.3	2
AMRAAM (Advanced Medium-Range Air-to-Air Missile)	12.4	10,917
AWACS (Airborne Warning and Control System) RSIP (E-3)	0.9	33
B-1 CMUP-COMP upgrade	0.6	103
B-1 CMUP-JDAM	0.9	95
C-130J	0.5	8
C-17A	42.2	120
DMSP (Defense Meteorological Satellite Program)	2.5	10
EELV (Evolved Expendable Launch Vehicle)	1.9	2
F-22	63.5	440
JASSM (Joint Air-to-Surface Standoff Missile)	0.7	44
JDAM (Joint Direct Attack Munition)	2.2	88,116

(Continued on next page)

TABLE 4A-1 *(continued)*

Weapon system	Total program cost	Quantity
JPATS (Joint Primary Aircraft Training System)	3.0	712
JSIPS (Joint Service Imagery Processing System)	0.8	35
JSTARS (Joint Surveillance & Target Attack Radar System)	9.8	20
MILSTAR satellites	. . .	6
MINUTEMAN III GRP	1.8	652
MINUTEMAN III PRP	2.3	607
NAS	0.7	53
NAVSTAR GPS satellite	8.9	118
NAVSTAR GPS user equipment	6.5	194,249
SBIRS (Space-Based Infrared System)	3.6	5
SFW (Sensor Fuzed Weapon)	1.9	5,084
Titan IV	23.1	41
Subtotal	196.0	. . .
Department of Defense		
CHEM DEMIL	13.3	18
JSF (Joint Strike Fighter)	20.8	n.a.
Patriot PAC-3	7.5	54
THAAD (Theater High-Altitude Air Defense)	7.7	n.a.
Subtotal	49.3	n.a.
Grand total	687.2	. . .

Source: Department of Defense, *SAR Program Acquisition Cost Summary, As of Date: June 30, 1997*, as reprinted in *Inside the Pentagon*, August 14, 1997, pp. 12–13.

n.a. Not available yet.

a. The QDR reduced some of these programs, but complete cost data were not available as this book went to press. Notably, the F-22 program was reduced from 438 to 339 planes, the F-18 E/F from 1,000 to between 548 and 785, the joint strike fighter from 2,978 to 2,852, the Marine Corps V-22 from 425 to 360, and the JSTARS program from 19 to 13 planes. Numbers in this table also sometimes differ from planned operational totals.

TABLE 4A-2. *Highlights of the 1998 Defense Appropriations Bill*

Costs in millions of dollars

Weapon system	Costs			
	Procurement	R&D	Total	Quantity
Strategic programs				
B-1B bomber modifications	114	...	114	...
B-2 bomber	331	356	687	...
Ballistic Missile Defense Organization (BMDO)	385	3,385	3,770	...
Service BMD				
Space-Based Infrared System (SBIRS)	...	214	214	0
Airborne laser	...	157	157	0
Joint aerostat	...	35	35	0
Fixed-wing aircraft				
F-22 fighter	75	2,077	2,152	...
F/A-18 E/F Super Hornet	2,192	244	2,436	20
AV-8B Harrier	302	11	313	12
F-15 Eagle	237	137	374	5
F-16 Falcon	82	...	82	3
Joint Strike Fighter (JSF)	...	946	946	0
F-14 modifications	287	12	299	...
C-17 Cargo Aircraft	2,180	111	2,291	9
Joint Primary Aircraft Training System	78	64	142	22
T-45 Goshawk	296	...	296	15
E-2C Hawkeye	324	65	389	4
V-22 Osprey	689	...	689	7
E-8 Joint Stars Aircraft	336	126	462	1
Ships				
SSN-23 Seawolf	153	68	221	...
New attack submarine (NSSN)	2,600	408	3,008	1
CVX future carrier	...	12	12	0
CVN-77 carrier	50	35	85	...
CVN-68 overhaul (USS *Nimitz*)	1,662	...	1,662	...
DDG-51 destroyer	43,569	148	3,717	4
LPD-17 amphibious ship	100	13	113	...
Arsenal ship	...	35	35	0

(Continued on next page)

TABLE 4A-2 *(continued)*

Weapon system	Costs			
	Procurement	R & D	Total	Quantity
Missiles				
Joint Air-to-Surface Standoff Missile (JASSM)/Joint Standoff Land Attack Missile (JSLAM)	. . .	176	176	0
Javelin antitank missile	202	8	210	1,274
AMRAAM air-to-air missile	164	49	213	273
Army Tactical Missile System (ATACMS)	98	. . .	98	153
Tomahawk	52	91	143	65
Joint Standoff Weapon (JSOW)	84	80	164	113
Sensor-Fuzed Weapon (SFW)	154	20	174	556
Joint Direct Attack Munition (JDAM)	95	37	132	3,341
Other systems				
OH-58D Kiowa Warrior helicopter	54	. . .	54	. . .
AH-64D Longbow Apache helicopter	512	. . .	512	. . .
UH-60 Blackhawk helicopter	297	. . .	297	28
RAH-66 Comanche helicopter	. . .	282	282	0
Bradley Fighting Vehicle upgrade	283	87	370	. . .
M1 Abrams tank upgrade	638	52	690	120
Crusader artillery system	. . .	324	324	0
Cooperative Engagement Capability Navy battle management	115	213	328	. . .
Milstar satellite	. . .	690	690	0
Predator Unmanned Aerial Vehicle (UAV)	141	. . .	141	15
Pioneer UAV	7	. . .	7	. . .
Endurance UAV	. . .	193	193	0
Tactical UAV	. . .	54	54	0
NAVSTAR GPS system	14	73	87	. . .
Evolved expendable launch vehicle	. . .	92	92	0

Sources: Center for Strategic and Budgetary Assessments, *Conference Agreement on the FY 1998 Defense Appropriation Bill,* September 30, 1997; *National Defense Authorization Act for Fiscal Year 1998,* H. Rept. 105-132, 105 Cong. 1 sess. (GPO, 1997); *Department of Defense Appropriation Bill, 1998,* S. Rept. 105-45, 105 Cong. 1 sess. (GPO, 1997); *Department of Defense Appropriations Bill, 1998,* H. Rept. 105-206, 105 Cong. 1 sess. (GPO, 1997); and *Making Appropriations for the Department of Defense for the Fiscal Year Ending September 30, 1998, and for Other Purposes,* H. Rept. 105-265, 105 Cong. 1 sess. (GPO, 1997).

5
MILITARY READINESS

Although not as riveting to most observers as issues of broad national strategy, war plans, or weapons procurement programs, military readiness is critical. Matters such as the quality and frequency of training, pay and other compensation, spare parts availability and equipment maintenance, recruitment and reenlistment rates, military personnel aptitude scores, and other determinants of near-term combat potential are just as important to the effectiveness of America's armed forces as the size of the force or the quality of weaponry.[1]

The main messages to come out of this chapter are that, while U.S. military readiness is much better than many critics allege, it is not without problems. Those problems will indeed require that operations and maintenance accounts and certain other specific parts of the Pentagon budget receive the large amounts of money envisioned for them in President Clinton's 1999 budget request. They also necessitate that the president's specific initiatives in the areas of spare parts acquisition, retention bonuses, family housing, health care, and

1. For two recent views, see Daryl G. Press, "Lessons from Ground Combat in the Gulf: The Impact of Training and Technology," and Stephen Biddle, "The Gulf War Debate *Redux*: Why Skill *and* Technology Are the Right Answer," *International Security*, vol. 22 (Fall 1997), pp. 137–46 and 163–74, respectively.

increased family support during deployments be enacted.[2] It also essential that the recommendations of the Quadrennial Defense Review (QDR) to free up nearly $4 billion in annual funding through more base closures and reductions in the Army National Guard be adopted.[3] There is no credible argument for keeping open excess bases, and only the weakest thread of an argument for keeping an Army National Guard combat force of the present size. The Congress is acting unwisely in opposing the Clinton administration on these policies.

These specific increases in resources are needed to arrest downward trends that have developed in certain measures of military readiness. The initiatives would have a net annual price tag of about $4 billion a year and ensure the high readiness of the U.S. military into the next decade. Thankfully, the needed funds appear in the administration's 1999 budget request for operations and maintenance. Even after subtracting out costs not directly related to readiness, such as environmental cleanup, health care, and certain contingency operations, that level of funding will keep real resources per troop above typical 1980s levels and allow for the initiatives recommended here to be carried out. It is important that those funds be appropriated by the Congress and sustained in the years ahead.

But this chapter is not simply an endorsement of the administration's readiness initiatives that appear in the 1999 budget request and a detailed program for how its added readiness funds should be spent. It also advocates a real pay raise for uniformed troops of about 5 percent, well above and beyond the rate of inflation and the 3.1 percent pay raise the administration is requesting in the 1999 budget submission. That increase would recognize the professionalism and diligence of today's U.S. military and help reverse recent downward trends in troop retention. The last sections of this chapter address two other matters. One refutes the charge often levied against peace operations that they strain the force unduly or weaken the military's "warrior spirit." The other considers the potential role of Army reservists in

2. For a brief summary of these initiatives, see *Budget of the United States Government, Fiscal Year 1999*, p. 139.

3. For a similar view, see Ike Skelton, "The Quadrennial Defense Review: Budgets," *Congressional Record*, daily. ed., May 5, 1997, p. H2163.

combat operations and argues for cuts even beyond those laid out in the QDR.

QUALITATIVE IMPRESSIONS OF READINESS

Before moving to specific quantitative measures of readiness, what do top military commanders generally say about the quality and readiness of today's U.S. armed forces?

Most views are positive. Admiral J. Paul Reason, commander in chief of the Atlantic Command, stated in 1997, "Returning to the fleet after two years in Washington, I see excellence at every level. In summary, the readiness of our fleet is high."[4] Similarly, Assistant Marine Corps commandant Richard Neal declared that same year that forward-deployed Marines were their readiest *ever*, and that general equipment readiness actually exceeded targeted levels.[5]

Not all officers feel similarly: the commander of the Navy's Third Fleet (in the eastern Pacific Ocean) describes current readiness as "satisfactory, but fragile." But one is hard-pressed to find more critical net assessments than that among senior uniformed leaders, at least among those who testified before Congress in 1997.[6] And in a farewell breakfast for defense reporters in August of that year, General John Shalikashvili, former chairman of the Joint Chiefs of Staff, acknowledged localized "cracks" in readiness but emphasized the generally high caliber and preparedness of today's U.S. armed forces.[7]

Most specific and objective measures of readiness discussed below confirm these overall positive impressions, as do most trends in commanders' overall measures of their individual units' readiness, known as C-ratings (see figure 5-1 for Air Force data). Today's high readiness is largely a result of the increased realism of all the services' training

---

4. Statement of J. Paul Reason before the House National Security Committee's Subcommittee on Readiness, March 3, 1997, p. 5.

5. Statement of General Richard Neal before the Senate Armed Services Committee's Subcommittee on Readiness, April 17, 1997, pp. 4–5.

6. Statement of Vice Admiral Herbert Browne before the Senate Armed Services Committee's Subcommittee on Readiness, April 17, 1997, p. 2.

7. Bryan Bender, "Top Officer See Readiness 'Cracks,' But No 'Epidemic,'" *Defense Daily*, August 29, 1997, p. 351.

FIGURE 5-1. *Air Force Readiness, 1987–97*[a]

Percent of reporting units

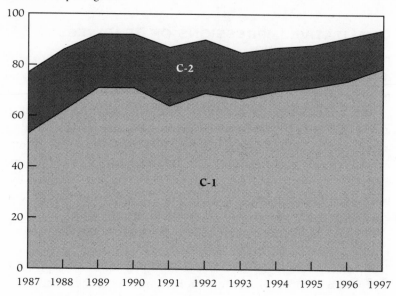

Sources: Department of the Air Force, *FY 98/99 Air Force Budget;* DoD Media Conference (February 1997); and Deborah Clay-Mendez, Richard L. Fernandez, and Amy Belasco, "Trends in Selected Indicators of Military Readiness, 1980 through 1993," Congressional Budget Office, March 1994, pp. 7–8.

a. Data as of January of each year. Units with C-1 ratings are deemed "fully ready" to carry out their wartime missions. Those designated C-2 are considered "substantially ready." The ratings cover personnel, training, equipment quantities, and equipment conditions. Each category is scored separately; the overall rating is usually the worst of the four scores.

regimens in the 1980s and 1990s compared with past decades. More recently, it is due to the fact that former defense secretary Perry kept overall readiness funding ample in recent years even when doing so required him to "rob" procurement accounts for several years' running. The proficiency and quality of today's military is reflected in the outstanding performance of U.S. troops in operations such as Vigilant Warrior in Kuwait and Saudi Arabia (1994), the would-be invasion of Haiti followed by the peacekeeping mission there (1994–96), maintaining no-fly-zones in Bosnia and Iraq, the implementation force (IFOR) and stabilization force (SFOR) operations in Bosnia (1995–

present), and the late 1997–early 1998 deployment to the Persian Gulf region. However, problems also are apparent. They require attention through a combination of careful force management and targeted increases in resources for personnel as well as operations and maintenance accounts.

FUNDING FOR READINESS

The crudest but most comprehensive measure of readiness is the per capita funding for operations and maintenance (O&M) of the forces, a major category of the Department of Defense's budget. It indicates the adequacy of resources for most elements that go into the near-term ability of units to carry out their missions promptly—maintenance of equipment, rigor and frequency of training, availability of spare parts and fuel, and the like. It does not cover funds for military salaries, however, and in aggregate form clearly does not show if O&M funds are being properly distributed and employed. But it is highly useful, nonetheless.

O&M per active-duty service member is now about $64,000 a year, in contrast to Reagan-buildup levels of roughly $51,000 (expressed in constant 1998 dollars). That overstates current readiness resources, however. When the added costs of health care, contingency operations, environmental cleanup, the drug war, Nunn-Lugar cooperative threat reduction aid to the former Soviet republics, and changes in accounting (some things like aircraft engines are now funded out of O&M, but were in the procurement account previously) are subtracted from the current figure to make the comparison more appropriate, today's figures look essentially like those of the Reagan buildup era.

Specifically, one should subtract from the 1997 spending level about $2.5 billion in health care, $0.5 billion in the drug war, $0.4 billion in Nunn-Lugar aid, roughly $3.5 billion in environmental cleanup, $3 billion in contingency operations, and $3 billion in accounting changes. A penalty of about $5 billion should also be subtracted for inefficiencies in a base, support, and reserve component structure that remains roughly 15 percent too big for the streamlined post–cold war military. These adjustments total $18 billion, thereby reducing the $64,000 per service member figure to slightly more than

$51,000—the average in the mid- to late 1980s in real or inflation-adjusted terms.[8]

What about plans for the future? A year ago, in the fiscal year 1998 budget request, the administration envisioned that unadjusted O&M budget per service member would be about $58,000. Assuming $2 billion less in costs for environmental cleanup (due to completion of the 1988, 1991, 1993, and 1995 base closure rounds), $2 billion less in contingency operations, and $1 billion less in waste from an excessive base structure, but some $2 billion more to carry out additional base closings that the Pentagon hopes Congress will have authorized by 2001, about $15 billion would remain in added O&M costs that were not present in the 1980s. Adjusted O&M spending per service member would drop to less than $47,000—well below the typical levels of the 1980s.[9]

Such a figure would have been too low, and the administration was wise to reconsider in the 1999 budget even though doing so required that future procurement accounts again be "raided" for the necessary funding. Planned budget levels for 1999–2002 now contain an average of $4 billion a year more in operations and maintenance funding, or some $3,000 per active-duty service member, than was projected for the same period a year before.[10]

READINESS OF EQUIPMENT

A key near-term readiness issue is equipment—specifically, its usability and dependability. (The technical caliber of equipment is usually

8. See Department of Defense, "FY 1998 Defense Budget Briefing," February 6, 1997; Ellen Breslin-Davidson, "Restructuring Military Medical Care," Congressional Budget Office, July 1995, p. 16; Deborah Clay-Mendez, Richard L. Fernandez, and Amy Belasco, "Trends in Selected Indicators of Military Readiness, 1980 through 1993," Congressional Budget Office, March 1994, p. 38; and Stephen Daggett and Keith Berner, "Items in the Department of Defense Budget That May Not Be Directly Related to Traditional Military Activities," Congressional Research Service memorandum, March 21, 1994, p. 39.

9. For a corroborating view, see Amy Belasco, *Paying for Military Readiness and Upkeep: Trends in Operation and Maintenance Spending* (Congressional Budget Office, September 1997), p. 5.

10. Department of Defense, "FY 1998 Defense Budget: Backup Charts," February 6, 1997; and Office of the Assistant Secretary of Defense (Public Affairs),

assessed separately by the Pentagon, and indeed is treated primarily in chapter 4 in this study.)

Consider depot maintenance—the major repairs done at central facilities, be they government-owned or private.[11] Certain guidelines exist by which the services allocate funds for conducting overhauls and repairs at such facilities, and the adequacy of those funds is often a good measure of how many ships, airplanes, and other systems will languish unusable while awaiting attention. In the 1980s the Navy typically funded depots at roughly 95 percent of anticipated aggregate needs, and in 1997 the level remained at 94 percent. Yet backlogs of Navy planes waiting for repair in depots have grown and now total about 170 units, with no immediate prospect for being reduced. Making matters worse, funding as a fraction of calculated requirements declined to 88 percent in 1998.[12]

Some equipment indicators remain acceptable but are nevertheless headed in an undesirable direction. This is true, according to Air Force Combat Commander General Richard Hawley, for such indices as mission-capable rates and "cannibalization rates" by which one aircraft is raided of some of its working parts to keep another flying.[13] Overall Air Force equipment mission-capable rates, though still comparable to 1980s levels, are now at their lowest in eight years.[14]

Increased funding for maintenance of certain specific weapons is needed to address some of the problems identified above. For example, the Air Force's C-135 transport fleet has a $50 million backlog that should be immediately funded. Certain other backlogs are

---

"Department of Defense Budget for FY 1999," February 2, 1998, p. 7. Planned procurement levels are down an average of $3 billion a year over the same period.

11. Deborah Clay-Mendez, *Public and Private Roles in Maintaining Military Equipment at the Depot Level* (Congressional Budget Office, July 1995), p. 2.

12. Clay-Mendez, Fernandez, and Belasco, "Trends in Selected Indicators of Military Readiness, 1980 through 1993," p. 43; and Office of Budget, Department of the Navy, *Highlights of the Department of the Navy FY 1998/FY 1999 Biennial Budget* (February 1997), pp. 2-5–2-8.

13. Floyd D. Spence, House Committee on National Security, "Military Readiness 1997: Rhetoric and Reality," April 9, 1997, p. 20.

14. Statement of Air Force General Thomas Moorman before the Senate Armed Services Committee's Subcommittee on Readiness, April 17, 1997, p. 5; and "Air Force's System-Wide Mission Capable Rate Drops to 76.6 Percent," *Inside the Air Force*, January 16, 1998, p. 4.

appearing, and some systems' mission-capable rates are declining. Even though their cause is less obviously a lack of funds, an increase in spare parts' funding of about 10 percent could go far to alleviate one major cause of these problems.[15] The U.S. Marine Corps has a similar problem; although it is funding spares at the historic "requirement" level, its backlogs are growing because its increasingly old equipment requires more expensive care. The Army is developing some problems with combat support and combat service support capabilities, including engineering and transport equipment.[16] On the whole, Department of Defense (DoD) equipment backlogs appear to be growing at about a half billion dollars a year and to have grown by about $3 billion since the end of the cold war.

Real net increases in funding for spare parts and depot repair of about $1 billion annually are necessary. They could reverse the recent trend and largely eliminate the maintenance backlog by 2002.[17] The recent boost given to operations and maintenance accounts in the 1999 budget request appears to provide the requisite funds.

READINESS OF TROOPS

The readiness of individual soldiers, sailors, Marines, and Air Force personnel is more complicated to assess, since it encompasses issues ranging from morale to experience to aptitude to commitment. It can be evaluated fairly well by looking at the training, accident and casualty rates, compensation levels, and retention and aptitude indices of today's troops and comparing them with previous eras.

Training

The rigor and frequency of training today remain quite good. For example, Navy "steaming days" for ships not on deployment remain at their

15. Statement of Air Force Major General George T. Stringer before the Senate Armed Services Committee's Subcommittee on Readiness, March 19, 1997, pp. 9–10.

16. Statement of Army Lieutenant General Thomas Schwartz before the Senate Armed Services Committee's Subcommittee on Readiness, April 17, 1997, p. 6.

17. Statement of Marine Corps Major General Thomas A. Braaten before the Senate Armed Services Committee's Subcommittee on Readiness, March 19, 1997,

FIGURE 5-2. *Navy Deployments, Fiscal Years 1971–99*

Steaming days per quarter

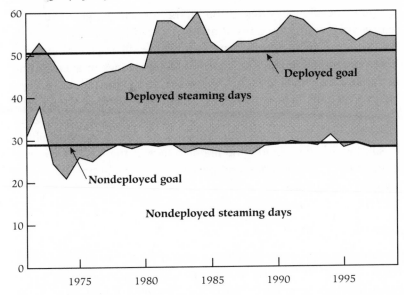

Source: Department of the Navy, Office of Budget, *Highlights of the Department of the Navy; FY 1998/FY 1999 Biennial Budget* (February 1997), p. 2-2.

targeted level of twenty-eight days per quarter, slightly higher than in the mid-1980s though recently reduced by one day per quarter as a result of improvements in the Navy's management of training operations (see figure 5-2).[18]

Other services' main training indicators also shape up well. Air Force fighter pilots train about nineteen hours a month, just as in the mid-1980s. Army tank mileage has dropped to about 650 miles a year, but is to be funded at the level of 800 miles a year in 1999, just shy of the mid-

---

pp. 2–4; Browne, statement, pp. 3–4; Assistant Secretary of the Army for Financial Management and Comptroller, "The Army Budget," April 1997, p. 71; and Wayne Glass, *Closing Military Bases: An Interim Assessment* (Congressional Budget Office, 1996), p. 27.

18. Office of Budget, Department of the Navy, *Highlights of the Department of the Navy FY 1998/FY 1999 Biennial Budget*, p. 2-2; and statement of Admiral Harold Gehman Jr. before the Senate Armed Services Committee's Subcommittee on Readiness, April 17, 1997, p. 3.

1980s level of about 850 (and perhaps compensated for through better simulation techniques).[19]

Some specific measures of the quality of training give cause for worry, particularly in units that have recently been involved in deployments. In some cases military skill levels or manning levels have deteriorated enough to cause notable declines in average unit performance at major military training centers, such as the Air Force's Nellis Weapons School and Army's National Training Center. The measures of decline are generally moderate in magnitude but real.[20] They arise at least as much from "traditional" security responsibilities in southwest Asia, notably the maintenance of the no-fly zone over Iraq, as from humanitarian and peacekeeping operations, but are of concern nonetheless.[21]

Operational Tempo

Deployment rates away from home, also often known as operational tempo (OPTEMPO), also remain too high for many individuals. The Army has been keeping many U.S.-based forces away from home 140 to 160 days a year—above the desired ceiling of 120 days. (These average figures are so high because they include time away for training and other purposes, as well as deployments.) Particularly for troops with families back at home base, and particularly in light of the fact that military personnel work hard even when at home base, this added strain above and beyond what servicemembers expect can be quite onerous.

The Air Force had similar problems a few years ago. It has since adopted rotation policies that distribute the deployment workload among more of the overall force structure, thereby limiting the number of people exceeding the 120-day ceiling to less than 10 percent of the force. Even so, some overdeployments continue, especially for RC-135

19. See Lane Pierrot, *Operation and Support Costs for the Department of Defense* (Congressional Budget Office, 1988), p. 37; Department of Defense, "FY 1998 Defense Budget Briefing," February 6, 1997; and Department of Defense, "FY 1999 Defense Budget Briefing," February 2, 1998.

20. Spence, "Military Readiness 1997," pp. 6, 15.

21. Statement of Air Force Brigadier General T. Michael Moseley before the House National Security Committee's Subcommittees on Readiness and on Military Personnel, March 4, 1997, pp. 2–3.

and HC-130 planes as well as U-2 and O/A-10 aircraft, and a growing number of systems are now being deployed at least 100 days a year—meaning that they are on the verge of overdeployment as the Air Force defines it. In this light, in addition to the Quadrennial Defense Review's suggestion to reduce joint exercises by at least 15 percent in the future and perhaps another 10 percent on top of that, the Air Force made the subsequent decision to postpone some of its large pilot competitions and training programs.[22]

Too high a rate of deployment also continues in many parts of the special forces. Generalizing across all special forces, enlisted personnel deployed about 120 days last year, whereas individuals in the most frequently needed units averaged 163 days.[23] Deployed ships' operational tempo remains somewhat higher than desired—about 55 days per quarter throughout the 1990s, in contrast to the goal of 50 days (see figure 5-2). But that level is typical of the last ten years and lower than in the early 1980s.[24]

A big part of the solution to excessive operations tempo is to ensure, as the Navy and Marines always have tried to do, that as many units as possible share in a given burden of work for which they are qualified. With this approach, the current pace of deployments should not be prohibitive. For example, the average of 14,000 Air Force personnel who were away from home base on mission during 1996, while a fourfold increase from late cold war levels, nevertheless represented less than 3 percent of the Air Force. The problem is admittedly worse when one views the percentage of the conventional combat force structure deployed overseas—between forces based in Iraq, Turkey, and the vicinity of Bosnia, for example, more than 10 percent of the tactical fighter force has been deployed at a

22. Moorman, statement, p. 3; John Hillen, "Superpowers Shouldn't Do Windows," *Orbis*, vol. 41 (Spring 1997), p. 244; Spence, "Military Readiness 1997," pp. 6–7; General Accounting Office, *Military Readiness: A Clear Policy Is Needed to Guide Management of Frequently Deployed Units*, NSIAD-96-105 (April 1996); and Philip Shenon, "Air Force Acts to Cut Stress among Pilots," *New York Times*, August 20, 1997, p. A20.

23. General Accounting Office, *Special Operations Forces: Opportunities to Preclude Overuse and Misuse*, NSIAD-97-85 (May 1997), p. 11.

24. Office of Budget, Department of the Navy, *Highlights of the Department of the Navy FY 1998/FY 1999 Biennial Budget*, p. 2-2; and Gehman, statement, p. 3.

time.[25] But even that hardly seems inconsistent with a policy that seeks to keep people at home two-thirds of the time.

The Air Force has ameliorated this problem by spreading the workload around more units. But it has learned by trial and error, and as a result has perhaps been slower to make adjustments than would have been desirable. For example, only in 1997 were unaccompanied tours in Saudi Arabia reduced from ninety to forty-five days, and only in 1998 is the burden of conducting them being spread across other parts of the force, such as units based in Japan.[26] With these modifications to rotation policy, the current operational tempo appears sustainable. Air Force pilot retirements are unsustainably high at present, however, suggesting that further pay incentives or operational tempo modifications may be needed in the future. Alternatively, the Air Force may simply need to plan on training—and ultimately losing to early retirement—more pilots on the assumption that retention will remain a serious challenge into the future (particularly for as long as the commercial airline industry keeps expanding).[27]

The Army's deployment burden was heavier in 1996, averaging 34,000 soldiers at a time (primarily as a consequence of the Bosnia operation). But it still represented less than 10 percent of that service's total manpower. That is a challenging pace, though one that should be sustainable in the end for a service that says it takes no more than four units in the force for every one that can be comfortably deployed.[28]

Indeed, the typical first-term Marine has been spending twice as much time away from home as an Army counterpart and five times as

25. Consistent with these data on deployed force structure, Alan Vick of the RAND Corporation has recently calculated that 10 percent of all Air Force flight hours from 1991 through 1995 were in support of contingency operations.

26. "Air Force to Spread Responsibility for Operations across Force," *Inside the Pentagon*, June 12, 1997, p. 1.

27. See David A. Fulghum, "Tough Schedules Thin USAF Ranks," *Aviation Week and Space Technology*, September 15, 1997, pp. 78–79.

28. Statement of Air Force Major General Donald L. Peterson before the House National Security Committee's Subcommittee on Readiness, March 11, 1997, p. 1; statement of Army Major General David Grange before the House National Security Committee's Subcommittee on Readiness, March 11, 1997, p. 4; and statement of John Shalikashvili before the House National Security Committee, February 12, 1997, p. 11.

much as an Air Force enlistee. The Marines averaged 23,000 individuals on deployment last year out of a force of just 174,000—in contrast to an Air Force total of 14,000 out of a force twice as large and an Army total of 34,000 out of a force nearly three times as large. Part of this higher tempo is a product of Marine Corps culture and the fact that its troops are generally younger and fewer are married than those of other services. But part is a function of the Marines doing a better job of sharing operational burdens among all their units and thereby avoiding excessive strain on individual elements. The Army and Air Force should try to emulate the Marines more in this respect.[29]

The operational tempo of troops could also be held in check by more discretion on the part of regional commanders than they sometimes display. There are ways to limit operational tempo without scaling back peace operations and other elements of U.S. efforts to "shape" the international security environment. As pointed out in recent congressional testimony by Admiral Reason, commander-in-chief of the Atlantic Command, regional commanders—the so-called CINCs, or commanders-in-chief—often do not have to "pay" for forces that they request to conduct exercises or operations. They make requests that the services try their utmost to honor, but at times fewer requests would be appropriate. Certain types of routine presence operations and exercises, in particular, can afford to be curtailed if the United States is already highly active in a given area and if such efforts are placing strains on the presence forces.[30] Recognizing this, the Quadrennial Defense Review has, as noted, appropriately called for a 15 percent reduction in "troop-days" associated with joint exercises in future years.[31]

This logic should not be pushed too far, since reliable presence is itself an important military element of U.S. global engagement policy. But it is not the only element and can be scaled back at times, particularly when other forms of U.S. defense involvement in a given part of the world have recently been manifest.

In the end, it must be recognized that operational tempo is a concern for today's U.S. armed forces, particularly in certain special-

29. Neal, statement, pp. 3–4.
30. Reason, statement, p. 4.
31. William S. Cohen, *Report of the Quadrennial Defense Review* (May 1997), p. 36.

purpose or special-capability units but even to a lesser extent in the core of the nation's combat capabilities. For that reason, the nearly 10 percent additional troop cuts advocated in this study, as part of a transition from a two–Desert Storm to a Desert Storm plus Desert Shield plus Bosnia structure, might best be made by reducing the size of individual units rather than their numbers. There is evidence that the Army is itself considering such a decision—and good reason to think that divisions have become too heavy of late.[32] However, in the case of Navy carrier battle groups and submarines, changes in deployment practices could permit cuts in actual force structure, as discussed in chapter 3, and even larger cuts in troops. In all, under my proposal the Army would lose 40,000 soldiers, the Navy almost 50,000 sailors, the Marines 13,000 uniformed troops, and the Air Force about 9,000 individuals; the U.S. military would decline from the QDR goal of 1.36 million active-duty personnel to 1.25 million. But the force would retain ten active Army and three active Marine divisions as well as twenty Air Force fighter wings.

Some actual enhancements to selected pockets of the force should be made. The Air Force has wisely bought more types of overused assets like RC-135 reconnaissance aircraft (two more), and increased crews for overtaxed systems like AWACS (eight more).[33] More should be done in this regard, not just for the Air Force but for Army units like military police, civil affairs and psychological operations specialists, and quartermaster and transportation units as well. For example, by adding about 7,000 individuals to the active-duty force structure, the Army could redress the shortfall it experienced in Somalia in such areas. Some other changes would require structural change but not added costs; for example, the Army could consider creating combined logistics companies that each contain a wide array of capabilities, rather than having individual specialized companies that are integrated only at the battalion level. Such smaller, modular support units would be better suited to the variable and generally modest demands of peace operations. All told, these types

32. Barbara Starr, "Each U.S. Army Division May Lose 3,000 Soldiers," *Jane's Defence Weekly*, November 5, 1997, p. 8; on Army heavy divisions, which have grown in weight by about 50 percent over the last decade, see Rachel Schmidt, *Moving U.S. Forces: Options for Strategic Mobility* (Congressional Budget Office, 1997), p. 82.

33. Fulghum, "Tough Schedules Thin USAF Ranks," p. 78.

of increases would add about half a billion dollars a year to DoD's budget.[34]

Morale, Recruiting, Retention, and Aptitude

What about the issues of military aptitude, motivation, and morale? Generally, the news here is good, though there are several causes for attention at present (see figures 5-3 through 5-5).[35]

Regarding reenlistment, actual rates are reasonably good, although recent trends are somewhat less reassuring. The effective rate at which existing military personnel chose to sign up for additional tours of duty was 65.5 percent in 1996, the lowest rate since 1990 but still at the average for the last fifteen years. (That effective rate is intended to exclude effects like forced discharges, which artificially lowered the actual reenlistment rate in the early 1990s.) The recent downturn in retention reflects as much the strength of the strong civilian economy as a lowering of morale among troops. But that does not ease the immediate problem. For example, Air Force pilots are accepting lucrative offers from commercial airlines, which typically pay more than twice their current salaries; as a result, the Air Force anticipates a pilot shortage of 360 individuals in 1998 and more than 700 in 1999 and 2000.[36]

Of greatest concern, the year 1996 also witnessed a fifteen-year low in the reenlistment rates of first-term troops. That recent downturn is particularly visible in the Army. Troops in Europe that bore the major brunt of the Bosnia operation have reenlisted at a higher rate than troops from other units, suggesting that extended deployments per se are not inconsistent with good retention. But total operational requirements and strains, together with opportunities in the civilian

34. Stephen T. Hosmer and others, "Bettering the Balance: Large Wars and Small Contingencies," RAND Issue Paper 167, 1997, pp. 3–4; and Peterson, statement, p. 2.

35. For a corroborating view, see the chapter by the former director of Pentagon Program Analysis and Evaluation in the Reagan and Bush administrations, David S. C. Chu, "What Can Likely Defense Budgets Sustain?" in Zalmay M. Khalilzad and David A. Ochmanek, Strategy and Defense Planning for the 21st Century (Santa Monica, Calif.: RAND Corporation, 1997), pp. 266–69.

36. Faye Bowers, "Air Force's Pressing Mission: Keep More Pilots on Board," Christian Science Monitor, September 19, 1997, p. 3; and Stringer, statement, p. 4.

FIGURE 5-3. *Total Department of Defense Reenlistment Rates, Fiscal Years 1982–96*[a]

Percent of total regulars

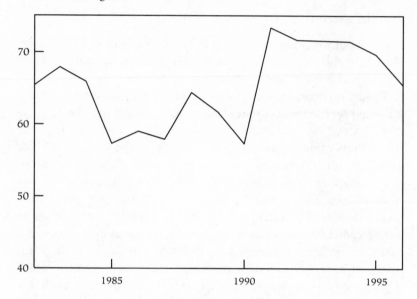

Source: Department of Defense, Washington Headquarters Services, Directorate for Information Operations and Reports, *Military Manpower Statistics*, 1996, pp. 34–37.
a. Adjusted rates, which exclude the effects of early separations and other discharges of eligibles under early release programs for strength control purposes.

economy and other factors, are now taking a toll on individuals' interest in making the Army a place for a long-term career.[37]

Other warning signs are becoming apparent as well. Two recent studies asking broad swaths of military personnel about their likelihood of making the military a career have shown a decline in those answering in the affirmative by at least 10 percent since 1991.[38] That year, just after the conclusion of Operation Desert Storm, reflected unusually high morale, so its use as a benchmark skews the result somewhat. But the point remains valid for relative trends over the last few years. Recent anecdotal evidence collected from frequently

37. Directorate for Information Operations and Reports, *Department of Defense Military Manpower Statistics* (September 30, 1996), pp. 34–41.

38. Spence, "Military Readiness 1997," pp. 19–20.

FIGURE 5-4. *Aptitude Scores of Enlisted Recruits, 1982–97*[a]

Percent of recruits scoring above average on the
Armed Forces Qualification Test

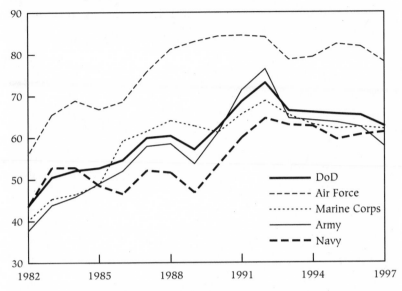

Sources: Department of Defense, Office of the Assistant Secretary of Defense (Force Management Policy), *Population Representation in the Military Services* (November 1996), p. D-14; and Department of Defense, Office of the Assistant Secretary of Defense (Public Affairs), "Fiscal Year 1997 Recruiting Meets or Exceeds All Objectives" (November 14, 1997).
a. Many individuals taking the test ultimately do not join the military, but their scores constitute a database against which one can evaluate those who are enlisted.

deploying units by the General Accounting Office also suggests some problems, though the data are of uneven quality and significance.[39]

With the post–cold war drawdown nearly complete, the Army now needs to increase its total numbers of recruits. Reflecting the fact that it expects to have difficulty doing so, the Army has lowered its goal for the percentage of new recruits with high school diplomas from 95 percent to 90 percent.[40]

39. General Accounting Office, *Military Readiness: A Clear Policy Is Needed to Guide Management of Frequently Deployed Units*, NSIAD-96-105 (April 1996), pp. 12–14.

40. Assistant Secretary of the Army for Financial Management and Comptroller, "The Army Budget," April 1997, p. 14.

FIGURE 5-5. *Age and Experience of U.S. Troops,*
Fiscal Years 1980–96

Mean age in years

Mean months in service

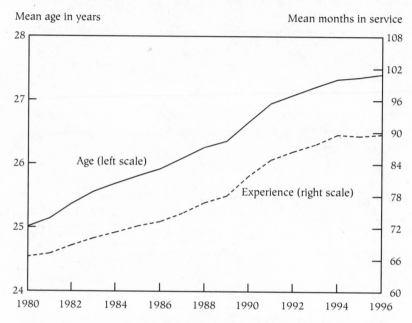

Source: Department of Defense, Office of the Assistant Secretary of Defense (Force Management Policy), *Population Representation in the Military Services,* p. D-17.

Even if troops are not particularly content in the armed forces, they are qualified. High school graduation rates among enlisted personnel were at a ten-year high in 1995 and improved further in 1996. Both two-year and four-year college completion rates among officers were their fourth best over that ten-year period.[41] Troops are also well disciplined. Court martial rates are lower now than in 1990 for all four services, though Marine Corps and Air Force numbers have taken a slight upturn in the last few years. Other forms of military punishment are also almost uniformly lower on a per capita basis, with the reas-

41. Shalikashvili, statement, p. 8; Department of Defense Washington Headquarters Services, Directorate for Information on Operations and Reports, *Department of Defense Selected Manpower Statistics, Fiscal Year 1995* (Government Printing Office, 1995), pp. 92–110; and Department of Defense, *Defense Almanac 96: Issue 5* (Alexandria, Va.: Armed Forces Information Service, 1996), p. 20.

suring downward trends continuing throughout the decade to date.[42] But these good indicators will be little comfort if reenlistment and recruiting trends continue in their recent directions.

Military Compensation

On the matter of general military compensation, the case for pay raises and other steps to reverse recent downturns in morale and retention is becoming rather strong.

Today's military compensation remains comparable to that in the private sector for jobs of comparable skill level. But it is not as good as during the Reagan years, in real terms, and may not be adequate recompense for the frequent deployments and time away from home that characterize today's U.S. military.

Pay has held its own against inflation throughout the 1990s (see figure 5-6). Current compensation levels are slightly greater than at the end of the cold war. But as noted, they are significantly—that is, roughly 7 percent—below the Reagan-era average.[43] Also, changes in the military pension program during this decade have reduced the expected receipts of a retiree with twenty years of service by 20 percent. Although sensible, such changes have changed the previous expectation of "deferred compensation" as a central element of the military benefit system.[44]

For these reasons, I recommend a one-time general pay raise for uniformed military personnel that would total 2.5 percent above the rate of inflation (and be followed by normal cost-of-living adjustments thereafter). That would respond to various difficult trends for the men and women of the U.S. armed forces of late. It would also make easier DoD's job of retaining high-caliber individuals in a highly competitive labor market. Such a raise would be especially welcome for young individuals with families, who often have a fair amount of trouble making ends meet.[45] This approach would

42. Dave Matfield, "Disciplinary Cases Drop in Armed Services," *Norfolk Virginian-Pilot*, June 29, 1997, p. 1.

43. See, for example, General Accounting Office, *Defense Budget: Trends in Active Military Personnel Compensation Accounts for 1990–1997*, NSIAD-96-183 (July 1996), pp. 3–12.

44. Shalikashvili, statement, p. 9.

45. This point is acknowledged by the chairman of the joint chiefs of staff. See ibid., p. 9.

FIGURE 5-6. *Military Compensation Costs, by Category, 1990 and 1997*[a]

Constant 1998 dollars per uniformed service member

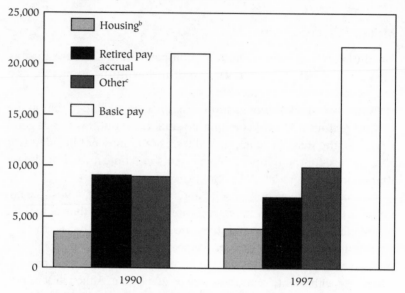

Source: General Accounting Office, Defense Budget, *Trends in Active Military Personnel Compensation Accounts for 1990–97*, NSIAD-96-183 (July 1996), p. 12.

a. Weighted average across all services.

b. Includes basic allowances plus a variable allowance for those in high-rent areas.

c. Allowances, special pays, and permanent change of station.

increase military pay by nearly $2 billion annually in the future—an expensive but worthwhile investment.[46]

Targeted pay increases or incentives for certain types of specialists also would make sense. For example, the Air Force is not only losing pilots, it is also having trouble recruiting other individuals with proper technical specialties. The Army is at risk of developing shortfalls in maintenance technicians, as is the Navy in regard to advanced elec-

46. DoD might take this opportunity to streamline or even eliminate its commissary and exchange system, using the $200 million or so in resulting annual savings to increase the size of the pay raise further. Some individuals would gain more than others from this change, and many would not like it, but it would on balance be desirable. It would get DoD out of a business the private sector is more competent to conduct and help communities by increasing the profitability of local businesses and the tax receipts received by local governments. See Deborah Clay-

tronics specialists.[47] Generally, increasing pay and incentives for individuals with the needed skills can be accomplished for tens of millions of dollars per category of speciality.[48]

DoD should also be more generous in its day care programs. Care should be expanded quickly and significantly, in contrast to the Pentagon's much more measured approach. Current overall provision rates are too low—only 60 percent of need in the Air Force, for example, and 65 percent in the Army. That is inadequate, particularly for a military that is often taking one parent away from his or her family on deployment (and generally not paying enough that the other parent can easily stay home full time).[49] Redressing this shortfall would cost about $200 million a year; it should be fully and promptly funded.[50]

Military housing is also being underfunded. Most is being provided by the private sector, as is appropriate, but in those places where a flexible and sizable private market does not exist, DoD continues to depend on its own facilities. Yet a $30 billion backlog of repairs is being funded at the rate of less than $1.5 billion a year, meaning that the backlog could endure well after 2010. Accelerating this schedule by one-third could be accomplished for roughly an additional $1 billion a year and would be a wise investment.[51]

CASUALTIES

Another broad measure, albeit a macabre one, of the strain on troops is the number who die from nonhostile causes in a given year. Higher

---

Mendez, *The Costs and Benefits of Retail Activities at Military Bases* (Congressional Budget Office, October 1997), p. xviii.

47. Schwartz, statement, p. 6; and Browne, statement, p. 4.

48. Statement of Air Force Brigadier General Steve Roser before the House National Security Committee's Readiness Subcommittee, March 3, 1997, p. 4; and Moorman, statement, p. 3.

49. Stringer, statement, p. 6; and statement by Major General Roger Thompson before the Senate Armed Services Committee's Subcommittee on Readiness, March 19, 1997, p. 6.

50. Assistant Secretary of the Army for Financial Management and Comptroller, p. 71; and statement of Assistant Secretary of Defense Fred Pang before the House National Security Committee's Morale, Welfare, and Recreation Panel, April 10, 1997, p. 8.

51. William S. Cohen, *Annual Report to the President and the Congress* (April 1997), p. 38.

death rates may reflect poor morale—particularly in cases of suicide. They also may reflect a lower caliber of training or worse condition of equipment being used in training.

Most trends in this area are, thankfully, for the better. Overall non-hostile deaths per 100,000 active-duty troops, which occurred at the rate of 125 a year around 1980, declined to about 70 by the end of the Bush administration. In 1994 they reached their fifteen-year low of roughly 65; in 1995 they numbered about 67, and in 1996 dipped backed downward again to roughly 1994 levels. In 1997 military aircraft losses were at an all-time low, and the flight accident rate of 1.5 per 100,000 flying hours was essentially unchanged by comparison with the previous two years and also very low by historical standards.[52] Suicides were at their low in the early 1980s, and homicides in the late 1980s, by comparison—but the overall numbers of losses are considerably below those from accidents (see figure 5-7).[53] A good way to continue this trend would be to expedite funding for aircraft collision-avoidance systems at a near-term cost of some $50 million to $100 million a year.

Despite the spate of military aircraft losses and ensuing casualties near the end of fiscal year 1997, safety trends were generally good for that year as well. For example, Navy aircraft loss rates were at an all-time low.[54]

OTHER READINESS ISSUES

A number of other ideas have recently been advanced that bear on the readiness debate. One school of thought claims that serious damage to the combat skills of U.S. troops may occur if they are involved for an extended period in operations other than war, such as peace-

52. Office of the Assistant Secretary of Defense for Public Affairs, news briefing, October 7, 1997, p. 1.

53. Department of Defense Directorate for Information on Operations and Reports, *Worldwide U.S. Active Duty Military Personnel Casualties* (1996), pp. 7–8. Over this period, 98 percent of all U.S. military deaths were from such nonhostile causes; a total of 539 individuals died from enemy action out of 29,929 to lose their lives overall—60 percent of whom died in accidents, some in training or operations, and others when driving or engaging in off-duty activities.

54. Steven Komarow, "Military's Air Arms Aim for Reassurance," *USA Today*, September 18, 1997, p. 1; and "F-14 on Training Run Crashes Off N.C. Coast," *Washington Post*, October 3, 1997, p. A28.

FIGURE 5-7. *Worldwide U.S. Active-Duty Military Deaths,*
by Cause, Fiscal Years 1980–97[a]

Nonhostile deaths per 100,000 troops

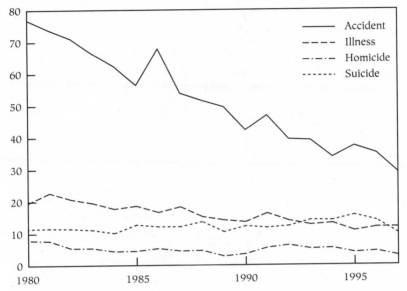

Source: Department of Defense, *Worldwide U.S. Active Duty Military Personnel Casualties:*
October 1979 through June 1997, Washington Headquarters Services, Directorate for Information Operations and Reports, 1997, pp. 1,7, 33.

a. There were 560 deaths from hostile action over the same period, including 256 in Lebanon,
18 in Grenada, 23 in Panama, 148 in Operation Desert Shield/Storm, and 29 in Somalia. The
total number of nonhostile deaths was 30,533. Data for 1997 are extrapolated, based on the first
nine months of the fiscal year.

keeping. Another involves Senator John McCain's proposal to save
money through "tiered readiness" in which some troops would be
kept more ready to go than others. A final idea suggests that Army
National Guard combat units are capable of playing greater roles in
U.S. war plans than they are now assigned. I do not find the case for
any of these propositions compelling, but all three issues are worth
considered attention.

Peacekeeping and Military Readiness

John Hillen of the Council on Foreign Relations argues that "superpowers shouldn't do windows." That is, because only the United

FIGURE 5-8. *United Nations Peacekeeping Operations Troop Levels and U.S. Contribution, 1990-97*[a]

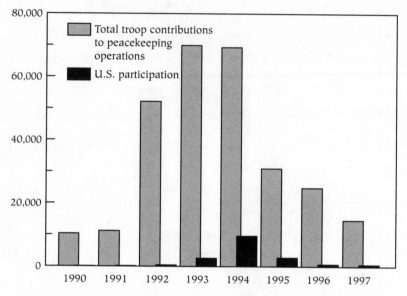

Troop levels

Source: United Nations, Department of Public Information, Peace and Security Section, June 1998.

a. Principal U.S. mission contributions (personnel): 1990: Middle East (33); 1991: Middle East (35), Western Sahara (29), Iraq/Kuwait (20); 1992: Former Yugoslavia (342), Cambodia (34), Western Sahara (27); 1993: Somalia (1,915), Former Yugoslavia (647); 1994: Former Yugoslavia (891); 1995: Haiti (2,226), Macedonia (565); 1996: Macedonia (489), Bosnia/Herzegovina (160). All data as of December 31 of each year, except for 1990, as of November 30.

States is capable of the most difficult military operations in the world today, it should allow others to handle missions that are less challenging.[55]

This policy has already been applied over the years to UN peacekeeping. The United States has never been a major troop contributor to such "blue helmet" operations (see figure 5-8).[56] It should continue

55. Hillen, "Superpowers Shouldn't Do Windows," pp. 241–57.

56. For example, the United States provided 818 out of a total of 75,000 peacekeepers in the field in 1994. See United Nations Department of Public Information, *United Nations Peace-keeping Information Notes* (December 1994), p. 242.

to avoid that realm, particularly for the safer types of monitoring activities that characterize peacekeeping in the limited sense of the world. But what of other types of missions?

As of 1997 the pace of such operations other than war appears sustainable for the U.S. military—especially if the services continue to improve their policies for troop rotations. But operations such as the IFOR mission in Bosnia placed a significant strain on the force and should be undertaken with particular discretion. Hillen's suggestion should be borne in mind carefully, particularly if the need for more than one such large mission arises at a time.

However, difficult peace operations often cannot be done without U.S. support. In some cases, U.S. political leadership is key; in others, U.S. power projection assets such as transport aircraft and ships, deployable logistics, and satellite communications are needed (to say nothing of standard combat skills, which are rivaled by few countries).[57]

Fortunately, the degradations to combat capability during such operations appear to be rather modest and remediable. According to Major General William L. Nash—who led the 1st Armored Division on its Bosnia deployment—they require about 100 days to redress. An even stronger vindication of the Bosnia mission's compatibility with unit readiness was offered by Colonel John Batiste of the 2d brigade of the 1st Armored Division, who said flatly that "the naysayers are wrong" in claiming that IFOR damaged combat capability. In particular, individual weapons skills were continually exercised and small-unit operations practiced throughout the mission.[58] Also, some types of skills such as engineering and logistics can actually be improved during such missions—as reported not only by the General Accounting Office, but by former Army general George Joulwan, who as supreme allied commander in Europe had overall responsibility for the Bosnia IFOR and SFOR deployments. Joulwan also pointed out the benefits to unit cohesion, discipline, and relations with other

57. On this point, see Michael O'Hanlon, *Saving Lives with Force: Military Criteria for Humanitarian Intervention* (Brookings, 1997).

58. Robert Holzer, "U.S. Army Units Stay Sharp in Bosnia," *Defense News,* June 2–8, 1997, p. 28.

countries, including Russia.[59] The main skills that suffer appear to be maneuver operations for larger units, like divisions.

Under a Desert Storm plus Desert Shield plus Bosnia warfighting strategy, a division actually engaged in peace operations would not be counted on to leave its mission in the event of war. And one that had just completed a peacekeeping deployment would also have ample time to train before undertaking a warfighting deployment. Other units that had not been involved recently in peacekeeping could be the first deployers, ensuring that a Desert Shield capability would be established in two theaters as quickly as possible. Given the deployment chronologies assumed in U.S. war plans, that approach should not pose a problem.

Large peacekeeping deployments—say, involving two or more divisions and a comparable number of air wings—could pose a greater concern. For one thing, in the event of a crisis, transport might not be available to take troops from their peacekeeping mission to training ranges where they could restore their warfighting skills. As long as the number of forces involved in operations other than war remains modest, this challenge can be handled.

Peacekeeping, peace enforcement, and other nontraditional military operations can also place serious strains on the military if dedicated funding is not made available for them. The Department of Defense has spent at least $10 billion over the last seven years on peacekeeping and related humanitarian operations (not including costs for operations in and around Iraq, which address more traditional national security concerns).[60] That funding level represents less than 1 percent of the Pentagon's budget. But it is larger than this

59. Statement of Army General George Joulwan before the House National Security Committee, March 19, 1997, pp. 12–14; and Grange, statement, p. 4. Some Army commanders advocate slightly longer than 100 days; for example, the Army's Center for Army Lessons Learned suggests a recovery period of four months. See Statement of Mark E. Gebicke, General Accounting Office, before the House National Security Committee's Subcommittee on Readiness, March 11, 1997, pp. 3–4.

60. General Accounting Office, *U.N. Peacekeeping: Issues Related to Effectiveness, Cost, and Reform*, NSIAD-97-139 (April 1997), pp. 3, 9; General Accounting Office, *Bosnia Peace Operation*, NSIAD-97-132 (May 1997), pp. 72–76; and William J. Perry, *Annual Report to the President and the Congress* (February 1995), p. 275.

country's UN peacekeeping dues (typically no more than $1 billion a year over that period). More to the point, it is significant relative to the accounts from which it must be drawn absent dedicated funding. Specifically, total Pentagon operating and training budgets total about $30 billion a year. They must fund a wide array of activities for a large and globally deployed military. There is little slack in these accounts (in fact, as argued above, they are insufficient in certain areas at present). In the end, these cost considerations are less an argument against peacekeeping than a caution. A president would generally be ill advised to undertake a peace operation without at least some degree of congressional support, since a lack of timely funding for the mission could seriously impinge upon the U.S. armed forces' overall readiness.

Tiered Readiness

In recent years some defense policymakers, led by Senator McCain, have argued that not all U.S. forces need remain in the same high state of immediate responsiveness for national crises. By their way of thinking, crisis-response forces such as those based overseas, the 82d airborne division, some special forces, and certain other parts of the force structure would be retained at high readiness levels. But money would be saved by allowing other units to remain in a less-ready state unless circumstances dictated otherwise. This idea would appear to have the greatest logic in regard to active-duty Army forces, many of which could not be deployed overseas immediately regardless of the severity of a crisis due to limitations on U.S. transportation assets.

In the Quadrennial Defense Review, the Pentagon examined the tiered readiness idea in some detail and found its promise wanting. The QDR's reasoning seems solid. The money that would be saved appears rather small—roughly $100 million a year for several Army divisions in aggregate—relative to the risks to quality inherent in any effort to assign certain units a second-class status.[61]

That QDR estimate appears a bit low. After all, the Army alone spends some $10 billion a year in operations and maintenance

---

61. Cohen, *Report of the Quadrennial Defense Review* (May 1997), p. 36; see also Chairman of the Joint Chiefs of Staff John Shalikashvili, "Report on Military Readiness Requirements of the Armed Forces," February 1997, pp. I-3–I-4.

accounts deemed to have at least limited flexibility. At least $2 billion of that appears directly tied to training and equipment maintenance.[62] It seems more plausible that the Army could save some significant fraction of that latter amount, perhaps half a billion dollars a year, through tiered readiness.[63]

In the larger scheme of things, however, $500 million is not a large amount of savings for such a major change to the way the force is structured and maintained. That is no surprise: tiered readiness does little or nothing to reduce most major costs of a unit, including the purchase price of its equipment, the salaries of its troops, the salaries of most of the civilians supporting its operations, and those basic training exercises it conducts. Only a fraction of the operations and maintenance account, itself only one-third of the defense budget, might be scaled back for such units, so it would be implausible to see savings much greater than 10 percent of overall cost.

Moreover, warfighting sharpness is best attained in a crisis by units that have already been very sharp at least once before. This approach would, by contrast, prevent some new soldiers from fully learning their trade when entering the service. They would have to await rotation to a first-tier unit before benefiting from the full panoply of military training opportunities. That seems a prescription for ensuring that the average soldier will not be as good in the future. The potential damage to morale and training that would result from specifically treating some active-duty units as second-tier is reason enough not to adopt the idea.

The Army National Guard Combat Forces

In today's U.S. military, Army reservists are more numerous than active-duty soldiers. That fact is in stark contrast to the situation in the other services, where the reserve component is in all cases less than half as big as the active component. Army Reserve and National Guard personnel represent almost two-thirds of all reservists and elicit the most debate in current defense policy. For that reason, they

62. Department of Defense, "FY 1998 Defense Budget Briefing," February 6, 1997.

63. For a concurring view, see Belasco, *Paying for Military Readiness and Upkeep*, p. xix. CBO's estimate is $450 million a year across the whole force.

merit the most analytical attention. They include nearly 370,000 National Guard personnel (mostly in combat units) and 215,000 Army Reserve personnel (mostly in support units for functions like logistics and military police). Their total annual budget of about $9 billion is nearly equal to that of the U.S. Marine Corps.

Although the Army reserve component is being asked to accept a relatively burdensome additional cut under the QDR, it is not suffering disproportionately over the whole 1990–2002 time frame. Moreover, if one takes a longer perspective, say 1980–2002, the Army reserve component is faring much better than the active forces.

Specifically, the QDR calls for a further 8 percent reduction in Army personnel in the Reserve and National Guard. That would take their aggregate size down from about 575,000 to 530,000. The reduction is larger than the QDR's 4 percent trim in active-duty forces but smaller than its 11 percent planned cut in civilian DoD employees. If completed as recommended, it will bring the total size of the Army reserve component down by 28 percent since 1990. Over the same period, active Army forces are being reduced by 36 percent.

In addition to becoming smaller, the Army reserve component is reorienting its mix of missions. Specifically, at least two divisions and additional brigades will be converted to much-needed support capability, reducing the Army's estimated shortfall in that area from 58,000 individuals to roughly 16,000.[64] That will leave the Guard with about 30 brigade-equivalents of combat units—down about 40 percent from the 1990 level (the active-duty Army reduced its divisional strength by 45 percent over the same period), but still leaving it with at least 100,000 more combat troops than war plans presently call for.[65]

To round out the comparative historical perspective, it is important to note that the Army Guard and Reserves grew greatly in the 1980s. In contrast, the active-duty Army grew little during that time. Thus Army reserve component end-strength will roughly return to its 1978 level under the QDR, if the Pentagon's recommendations are accepted

---

64. Frances Lussier, "An Analysis of the Army's Force Structure: Summary," CBO Memorandum, April 1997, p. 11.

65. Frances Lussier, *Structuring the Active and Reserve Army for the 21st Century* (Congressional Budget Office, December 1997), p. 11.

by Congress, whereas active-duty strength will be down more than one-third compared with that period.[66]

Reasons to Cut Army National Guard Combat Forces

It is unfair to criticize Guard combat brigades and divisions for the fact that they were not sent to Desert Storm. That decision was largely based not on their readiness for combat but on the desire of national leaders to limit the total magnitude of the reserve call-up. However, the weight of evidence still argues that the QDR was right not to factor most National Guard combat units into current war plans and that today's U.S. military has considerably more than it needs.[67]

First, some common arguments in favor of Guard combat units deserve rebuttal. It is often said that the reserves play an important role linking the American people to its armed forces. That is true to an extent. But as Army Lieutenant Colonel Michael Meese and Captain James Schenck of West Point recently argued in the *Washington Times*, the bond with the public "will not be enhanced if the National Guard and Army Reserve are seen as welfare entitlements that serve little national purpose." It is also often claimed that National Guard combat units cost just 25 percent as much as active-duty troops. That figure, however, relates only to operations and support costs—about 60 percent of the total for active-duty forces. To be truly as good as their active counterparts, Guard forces would need as good equipment (otherwise, why does the country spend its money on high-caliber gear for any forces?). With that approach, they would be more than 50 percent as expensive as active-duty troops—still a large savings, but less than frequently advertised.

The best argument against depending heavily on reserve component combat forces, however, is that the entire post-Vietnam history of the U.S. military underscores the importance of high readiness.

66. Michael Meese and James Schenck, "Factors in the Reserve and Guard Cuts," *Washington Times*, July 6, 1997, p. B4; Dick Cheney, *Annual Report to the President and the Congress* (Department of Defense, January 1990), p. 76; statement of Richard Davis, Director, National Security Analysis, National Security and International Affairs Division, General Accounting Office, before the Subcommittee on Readiness, Committee on Armed Services, U.S. Senate, "DoD Reserve Components," NSIAD-96-130, March 21, 1996, p. 1.

67. Cohen, *Report of the Quadrennial Defense Review*, p. 32.

Warfighting and other military operations are difficult and require diligent training and preparation, as emphasized in the Army's history of the Gulf War, *Certain Victory*.

When conducting their training, reservists may well be as diligent and dedicated as active-duty forces. Also, they may often be as qualified at their own specific technical tasks, especially in cases where their civilian jobs or other activities help keep them practiced on related skills. It is for this reason that the U.S. military total force concept makes such good sense. So does the Army's recent initiative to convert twelve Army National Guard combat brigades to support and service support units to redress shortfalls in those areas (with some of the needed funds now to come from reductions in reserve personnel advocated by the QDR).[68]

Several weeks of total time per year does not, however, appear sufficient to keep a combat unit sharp, particularly a large unit like a division. Some smaller units, such as those employed with the Marine Corps force structure, appear able to quickly get up to speed in the event of a national emergency, as evidenced in Desert Storm. But in the case of the Marines, that philosophy translates into a reserve component only one-fourth the size of the active force.[69]

Because of the need to minimize the amount of territory lost early in any future war against a regional foe, current U.S. war plans correctly call for deploying all ground forces to a major theater war quickly. Specifically, they call for a five-division force to be fully in place within roughly seventy-five days of a decision to deploy and for roughly half that force to arrive in theater within the first month of the crisis. That means, in turn, that the last units would set sail for their destination after about fifty days.[70] My view is that the goal should be to deploy even faster and that DoD should purchase more fast sealift and airlift and pre-position more materiel overseas to accomplish that goal.

There is good reason to think that DoD cannot meet its own goal today.[71] But it already has several remedial programs in place. In any

---

68. Ibid., pp. 33, 47.

69. Lane Pierrot, *Structuring U.S. Forces after the Cold War: Costs and Effects of Increased Reliance on the Reserves* (Congressional Budget Office, September 1992), pp. 14–20.

70. Schmidt, *Moving U.S. Forces*, p. 79.

71. Lussier, "An Analysis of the Army's Force Structure: Summary," p. 9.

case, the proper way to solve this problem is not to lower the goalpost (potentially allowing Guard units to be ready by the time lift would be available to carry them) but to buy more lift. Only a few days of warning, or even less, might be provided by the North Koreans or Iraqis should they elect to attack their neighbors again; in this geostrategic environment, rapid U.S. responsiveness is indispensable.

That type of emphasis on speed, which is likely to increase further in the future under most DoD long-term vision statements, spells trouble for a high dependence on reserve combat units. Various estimates claim that the Army National Guard's enhanced separate brigades would take 30 to 90 days to get ready and that divisions could take two months to a year. The Army's official position is at the higher end of those ranges.[72] A thorough RAND study suggests that, even with enhancements to its training capabilities and ample supplies of spare parts and ammunition that might be needed in the combat theater instead, the Army would need 102 days to ready three heavy enhanced brigades for combat and another 54 days to get three more ready. (Of the fifteen enhanced brigades, seven are heavy, seven are light, and the remaining unit is an armored cavalry regiment.)[73]

A recent study by the Institute for Defense Analyses (IDA), though not publicly available, reportedly argued that the 49th Armored Division of the Texas Army National Guard could validate its readiness 94 days after call-up and be deployed to its destination by day 132 (presumably including about 30 days of sailing time after 7 to 10 days to reach its port of debarkation).[74] In the absence of convincing documentation, there is little reason to have great faith in this report. But even if the relatively optimistic IDA estimate is correct, that would have the reserve division arriving in theater two months behind the schedule called for in current war plans for the main fighting force. Worse yet, due to the need to rotate units through the national training centers before validation of their readiness, the second and subsequent divisions would arrive much more slowly than that. Another recent RAND study, for example, suggested that under

72. Pierrot, *Structuring U.S. Forces after the Cold War*, p. 19.

73. Thomas F. Lippiatt and others, *Postmobilization Training Resource Requirements: Army National Guard Heavy Enhanced Brigades* (Santa Monica, Calif.: RAND Corporation, 1996), pp. xv–xviii, 1–21.

74. See Francis Greenlief, "Structuring the Total Force," *Defense News*, July 21–27, 1997, p. 37.

current conditions it would take 70 more days to prepare six enhanced separate brigades than to prepare just one.[75]

It is precisely because keeping forces combat-ready is difficult that the Congress has rightfully kept a careful watch on the strains and disruptions that operations other than war (OOTW) cause active-duty troops. Although generally a supporter of such OOTW, I fully concur with the need to ensure that the combat readiness of most U.S. units is not measurably impaired in the process. There is a legitimate debate over whether troops that have been on OOTW need three months, four months, or even longer to recover from the operation and regain full combat competence. But no one claims that they can be restored to combat readiness in less than three months. Common sense suggests that reservists would require at least as much training time after a mobilization, and probably more.

In this light, the case for increasing the nation's warfighting dependence on Army National Guard combat units appears weak. Some wish to keep all the Guard combat units as insurance anyway. But the fifteen enhanced separate brigades already have the role of providing insurance. They are not all needed, according to present war plans. The Total Army Analysis 2003, which actually is a slightly more conservative analysis than the BUR and QDR and allocates more Army forces to possible regional wars than they do, assumes that only 30,000 of the 175,000 Guard combat troops would participate in a full-scale two-war scenario of the type that drives current defense planning and is assumed to require all active-duty ground forces. That number of Guard forces could be provided by six of the enhanced brigades; the other nine would remain as backup in case things went very badly.[76]

In my view, the QDR overestimates somewhat the threats posed by Iraq and North Korea and understates the strength of South Korea as well as of U.S. airpower and equipment already pre-positioned in

75. Lussier, "An Analysis of the Army's Force Structure: Summary," pp. 23, 29.

76. The Total Army Analysis assumes that 195,000 Army combat troops would be needed for two major theater wars (that figure does not include support troops, which would number 477,000 more). Of that total, 165,000 would be provided by the active Army, including roughly 5 1/3 divisions for each war, or 32 of the active Army's 33 combat brigades in all. Lussier, "An Analysis of the Army's Force Structure: Summary," pp. 7–8; and Lussier, *Structuring the Active and Reserve Army*, pp. 7–11.

northeast and southwest Asia. Under these circumstances, even more insurance against these regional wars seems a poor use of the nation's defense dollar. Paying for more combat-oriented reservists than war plans require will worsen what is already a serious DoD budget problem and detract from the nation's ability to purchase needed equipment and ensure troop readiness.

Reasons Not to Cut Too Much

Having said all these things, which might lead one to wonder why the nation should keep any Guard combat units, several reasons for retaining the fifteen enhanced separate brigades and at least a half dozen additional brigades deserve mention:

—As the experience of the Marine Corps in Desert Storm demonstrates, smaller ground-combat reserve units can be highly effective after a short postmobilization training. The evidence seems best for units at the company and battalion level, but may apply to brigades as well, as a 1992 CBO study by Lane Pierrot noted.

—These smaller reserve units may be able to "backfill" units from active divisions that are called upon to conduct peace operations, allowing a division that is partially deployed for such a mission to reconstitute more quickly should a larger combat scenario develop.[77]

—If and when the nation is able to relax its vigilance toward northeast or southwest Asia, the need for current numbers of active-duty forces may no longer exist. At that point, faced with budget constraints and a generally peaceful but still potentially dangerous international environment, the nation may elect to give the National Guard proportionately more of the nation's ground-combat responsibilities. That would allow the Pentagon to save money, using scarce funds to maintain high readiness for a smaller active force, fund operations other than war when needed, and develop new technologies and new capabilities, such as missile defenses.

—In the short term, if the Army National Guard's fifteen enhanced separate brigades prove their mettle, there could be an argument for increasing their role in the nation's formal war plans and even

77. Stephen T. Hosmer and others, "Bettering the Balance: Large Wars and Small Contingencies," RAND Issue Paper 167, 1997, p. 4.

increasing their number slightly while decreasing slightly the size of the active Army.

—Deepening concerns over terrorism, perhaps including the use of chemical or biological agents, attacks against key nodes of the nation's information and industrial grid, or other security challenges within the United States, may increase the relevance of the Guard's territorial-defense mission in the future, as the secretary of defense, chief of staff of the Army, and National Defense Panel recently argued.[78]

These arguments should not be pushed too far, however. Future demands on U.S. forces are likely to be high even without, for example, an acute North Korean threat.[79] An active-duty force of more than 1 million will probably be needed simply to maintain global presence, conduct joint exercises, deter at least one major theater war, and be prepared for the unexpected. The Army will likely need considerably more than 300,000 active-duty troops even in such a world.

Also, as new technologies and tactics are developed in the future, a large Guard with generally obsolescent weaponry will do the nation little good. Factoring in the costs of equipment, the Guard is hardly cheap, as noted earlier, and should not be thought of as some great repository of latent national might. That latent national might will exist much more in the form of the nation's technology and industrial base and in the caliber of its active-duty forces that would serve as the nucleus of any major defense buildup and mobilization.

In short, the Army National Guard combat forces constitute a valuable national asset but one that is not great enough to justify even the QDR-recommended force of roughly thirty brigade equivalents. In an era of considerable fiscal stringency, a massive Guard combat structure is a luxury the Pentagon cannot afford.

CONCLUSION

The post–cold war defense drawdown is nearly over, as indeed it should be. U.S. forces are roughly one-third smaller and less expen-

78. George Seffers, "Guard May Fight Terrorism," *Defense News*, September 15–21, 1997, p. 18.

79. Richard L. Kugler, "Managing Regional Security: Toward a New U.S. Military Posture Overseas," in Khalilzad and Ochmanek, *Strategy and Defense Planning for the 21st Century*, pp. 213–54.

sive than in 1990, and the recent Quadrennial Defense Review will adjust them only at the margins. Defense spending will soon drop to 80 percent of its average 1948–90 real-dollar level and represent a share of the national economy (less than 3 percent of GDP) that has not been so small since before World War II.

Nevertheless, the United States still possesses by far the best military in the world and a force that matches up well against its predecessors of the 1980s and early 1990s. It is proving its mettle this decade across a remarkable variety of missions, ranging from full-out combat to peacekeeping to deterrence to joint exercises and other forms of "engagement" with allies and neutrals alike. Its operations are organized within a strategic framework that properly avoids simple characterization and responds to a broad set of challenges in a world that remains dangerous—even if much improved from its cold war nature. For all these accomplishments, the men and women of the U.S. armed forces and the last several administrations and Congresses can be very proud.

But challenges remain. Most notably, the U.S. military remains a bit too expensive in light of the fiscal environment and the resources likely to be made available to it in peacetime. The funding shortfall could soon translate into deterioration of the force and constraints on its ability to carry out various missions in the world.

By asking the major allies to help more, but even more by making a number of further judicious decisions about how to structure the U.S. armed forces and expend resources at the Pentagon, this problem can be solved without prejudice to U.S. global interests or leadership. The sooner we start thinking about this challenge the better. More tough defense decisions lie just around the century's corner—if not even closer.

INDEX